D1094864

AGRICULTURE AND MODERN TECHNOLOGY

THOMAS R. DEGREGORI earned his Ph.D. in economics from the University of Texas. He is a professor of economics at the University of Houston specializing in science and technology for economic development. He has a vast range of publications and has lectured and consulted worldwide.

© 2001 Iowa State University Press
All rights reserved

Iowa State University Press
2121 South State Avenue, Ames, Iowa 50014

Orders:	1-800-862-6657
Office:	1-515-292-0140
Fax:	1-515-292-3348
Web site (secure):	www.isupress.com

Authorization to photocopy items for internal or personal use, or the internal or personal use of specific clients, is granted by Iowa State University Press, provided that the base fee of $.10 per copy is paid directly to the Copyright Clearance Center, 222 Rosewood Drive, Danvers, MA 01923. For those organizations that have been granted a photocopy license by CCC, a separate system of payments has been arranged. The fee code for users of the Transactional Reporting Service is 0-8138-0342-X/2001 $.10.

♾ Printed on acid-free paper in the United States of America

First edition, 2001

Library of Congress Cataloging-in-Publication Data
DeGregori, Thomas R.
  Agriculture and modern technology: a defense / Thomas R. DeGregori—1st ed.
    p. cm.
  Includes bibliographical references and index (p.).
  ISBN 0-8138-0342-X
    1. Technology—Social aspects. 2. Technology—Philosophy.
  3. Agricultural innovations. 4. Agriculture—Environmental aspects. I. Title.

T14.5.D455 2001
303.48′3—dc21   00-054147

The last digit is the print number: 9 8 7 6 5 4 3 2 1

AGRICULTURE AND MODERN TECHNOLOGY

A DEFENSE

THOMAS R.
DEGREGORI

Iowa State University Press / Ames

Contents

Dedication

To
Gayle
Alice, James, Roger

Preface

This book deals with one of the great paradoxes of the twentieth century. It was a century characterized by economic and technological gains of unprecedented rapidity as shown by all of our economic indicators; an increasing number of economists now argue that they substantially understate the magnitude of these gains (Nordhaus 1997; DeLong 1991–2000). The noneconomic indicators are just as spectacular if one looks at life expectancy, health, and the increase in per capita food supply that more than accommodates an increase in population that virtually all "experts" believed could not be fed. Agriculture had the additional challenge of putting nutrients into the soil to feed the plants so crops could be grown to feed the growing population. The paradox is that these gains are largely denied by some groups, if only by implication, and the science and technology that allowed them to happen have been under attack for almost the entire century. The alleged dangers of modern life have become conventional wisdom to large segments of the population. I believe the attack is a manifestation of an antitechnology elitism.

In the opening chapters of this book, I argue that we are inherently technological beings and that the biological evolution that made us human was inextricably bound with the evolution of our early technology. If technology is in some sense an integral part of our being, it is difficult to see how it is also alienating us. Technology has been with us as long as we have been human. Antitechnology elitism has a lineage in Western culture going back to the Greeks, and there is a similar antitechnology elitism in other cultures, such as that of China. My early chapters explore the development of humans as technological beings and the concept of alienation as applied to our use of technology.

The core chapters concern the issues involved in the contradiction between the past century's gains in the human condition and life expectancy and the widespread belief that modern technology and science are uniquely life threatening. Are "chemicals" killing us? I ask the question in the text and elsewhere. If modern technology and "chemicals" are killing us, why are we living so long? I ask other questions as well: Is our food and water supply destroying the environment? Is modern medical practice deficient? Are there alternatives to these essential products of modern technology and science that are safe, "natural," and environmentally benign?

My closing chapters pick up the scientific issues of earlier chapters and explore their implications for policy actions. The intent of this book is not simply academic in the pejorative use of that term. I argue that the antitechnology, antirational views that have gained such a stronghold in many areas of academia and other elite groups, no matter how benign they may sound, have adverse consequences and harm some of the world's most vulnerable peoples. Critical in these chapters is the debate over the level of certainty necessary in order to justify a policy of using a technology or banning one. To the assurance of certainty demanded by many antitechnology zealots, I counter with John Dewey's idea of a warranted assertion. In the final chapter, I sum up what I consider to be overwhelming data on the benefits of modern technology using health and demographic data.

Conventional wisdom has one set of answers and this book has answers that are almost exactly contrary. For example, an environmentalist survey research organization ran ads in major U.S. newspapers claiming that 90 percent or more of Americans believed that environmental "chemicals," pollution, and waste are major causal agents for health problems such as asthma and various chronic diseases (Whelan 2000). I am part of the 10 percent that doesn't believe this. When I sent out the prospectus for this book, I stated that it was politically incorrect and controversial. One response suggested that this was an understatement.

Many of the ideas that I offer are more than controversial. Otherwise well-educated people in many cases are unaware either that these ideas exist or that there is a substantial body of scientific knowledge supporting them. Given the controversial nature of the ideas presented, I have generously documented each issue, with a purpose in mind. Original sources are at the core of the arguments in support of every major and most every minor issue. To these I have added material from secondary sources. Due to the vast amount of documentation resulting from my research, I present the essential sources here and have posted the others on my homepage (http://www.uh.edu/~trdegreg), by page and then by topic. I invite the reader to check out these sources, since most of them can be found online.

This book could not have been written without the internet and links from posted news articles to original sources. Being able to go from a news article to the peer-reviewed scientific study offered an incredible array of diverse sources, including access to peer-reviewed specialist journals. Equally important to my research have been the many news groups that send out daily dispatches via the internet with various useful writings, summaries, and hyperlinks to websites, magazines, journals, and newspapers around the world. They have also been extremely useful in allowing me to post questions on important issues and receive replies from leading scholars around the world. Observing the vigorous give-and-take of ideas and their advocates is exciting and as informative as eventually reading the articles that were later published. Happily, most of those in online discussion groups have curious minds open to new ideas. Unfortunately, the internet can also be a haven for closed minds seeking confirmation of their fervently held preconceptions. In sum, the electronic information media facilitate an exposure to an extraordinarily broad wealth of ideas and information that provide a worthy context and framework for ideas that one chooses to pursue in greater depth.

Among the many news groups that I receive, two in particular have been helpful, Agnet by Douglas Powell and AgBioView by C. S. Prakash. Specialized websites such as Junkscience.com by Steve Milloy have also been helpful. They are responsible for many of my sources and I express my thanks to them.

Using the internet for research has its advantages and disadvantages. The traditional printed method of referencing does not always mesh with electronic technology. When one quotes from an online posting of a printed article, it is generally not possible to give the page number. For those who go online to check sources, this is no problem since they simply search the article using a couple of words from the quote. Those who go for the hard copy may have some difficulty finding the quote and its context. With a few exceptions, including one of my own where the cited article is available only online, my bibliography follows tradition, refers to the hard copy, and does not provide the URL (which for some of my regular sources changed over the course of my research). In most cases, a search engine will find the journal or other source for those who prefer to go online as quickly as one can find the library call number.

There have been times when I pursued another author's sources and found errors in the documentation. I have always favored a complete reference, and in this book I have made my references as complete as possible. The help of the University of Houston library reference desk was invaluable in achieving this goal. In my judgment, my references are more than adequate for the reader to test, verify, or falsify the ideas expressed here.

All endnotes contain additional source material directly related to the text or to closely related points of interest to the reader. The endnotes contain little other substantive material. This allows the reader to continue through the text without interruption except possibly to make a note to check an endnote for additional sources after finishing the chapter.

One further note regarding sources: in citing newspapers, I used the date on the edition that I was reading, which means that the article may well have appeared on a different date in another edition if the newspaper had a wide publication area. This is normally only one day off, but could be more when international editions are involved. This is not a problem if one searches online, but it might be if one is searching an archival newspaper or microfilm.

In the ten years that I have worked on this book, many people have helped in all manner of ways. To all those who helped, please accept my most heartfelt thanks. My thanks also to the University of Houston library staff in the Interlibrary Loan and Reference Departments.

There were those whose consistent help stands out. Dr. Mihaela Y. Smith of Commonwealth Partnership for Technology Management has for the last decade and a half been making copies of my books and articles available to anyone who passed through her London office. On some occasions, she took copies on her visits to Africa and Asia, leaving them with government officials. The feedback was extremely valuable.

It is my good fortune that the Economics Department shares the same floor with the Anthropology Department at the University of Houston. At one point or another, I have discussed the ideas presented here with virtually all of the faculty in these two departments. My anthropology colleagues Rebecca Storey and Randolph Widmer were a seemingly endless source of references on the anthropological, demographic, and medical issues. My wife, Gayle, and my anthropologist sister-in-law, Anne Sutherland, have discussed many of the ideas expressed here as well. Gayle provided early editing.

This book, then, is truly a collective effort of far more people than I am able to thank. Let me add the two anonymous referees whose very careful and thoughtful review and criticisms allowed me to improve the work greatly. However many contributed, as is standard, eminently right, and appropriate here, I take the blame for all errors and omissions.

AGRICULTURE AND MODERN TECHNOLOGY

CHAPTER 1

Technology and the Unity of Experience

Humans strive to realize limitless potential. Unfortunately, at various stages of the development of civilization, higher achievement for some has been bought at the cost of the degradation of others. Being associated with technology was not always highly regarded. In fact, people involved with the practical arts and vocations of technology have been viewed, in the past, as inferior.

In Western civilization, ancient Athens is noted for the contempt of its philosophers for those whose pursuits were other than ethereal. Aristotle would have denied them the rights of citizenship (Aristotle 1958, 107–9). Plato would have denied citizenship to peasants, artisans, and mechanics and would not have wanted his daughter to marry one of their sons (Plato 1960, 234–35; Plato 1952, 89). The attitude toward those who worked furnaces that made the metals for Iron Age Greece was one of even greater scorn. Xenophon said those who tended the fires carried "a social stigma" and were "rightly dishonored" in the Greek cities (Farrington 1944, 28).

There were two reasons the Greek philosophers felt the way they did. Mechanical arts were unpleasant to perform and debilitating to those who carried them out. Working with fire meant "blood, toil, tears, and sweat" (Farrington 1944, 26). Given their unpleasantness, such tasks were carried out by slaves. The stigma of this association carried over to free men engaged in manual technological endeavors. No wonder that, in Greece and elsewhere, higher pursuits were preferred by those who were able to follow them.

In India at the turn of the twentieth century, scholars blamed "the estrangement of the hand from the mind for the decline of science"(Ghosh 1994, 6; Ray 1902–3, 1909). Alsop (1982) says that in the early centuries of Chinese civilization, "any work of art smelling too much of technical training, including technically skillful painting, tended to be looked down by many Chinese art theorists as being too close to artisans' work. On the other hand, a proper Confucian gentleman could not conceivably work with his hands—except to use a writing brush" (Alsop 1982, 223). Thus, for many hundreds of years, there was a strange dichotomy as the "incredible fertility of Chinese art was simultaneously enjoyed and disdained by serious Chinese art lovers" (223). Alsop notes, "The marvelous porcelains, the magnificent lacquers, the wonderful metalwork in gold, silver, and bronze, even the incomparable architecture and the vigorous sculpture—all these were thought unworthy of discussion by the numerous Chinese writers on art until a very late date" (223).

The consequence of this snobbery was that only the works of famous calligraphers and painters were "collector's prizes" for close to a thousand years. The brush that was used by the painters and calligraphers was also the tool of the scholars and therefore acceptable to them. "So it was both easy and natural for scholar-officials with a bent that way to begin using their skill with the writing brush to create paintings." Painting was then no longer "mere artisan's work" and could become accepted as a major art, a transformation that Alsop believes was paralleled in the rest of art history (Alsop 1982, 222, 223).

In their disdain for technology, the elites of various cultures have failed to recognize that it was past technological change that created the conditions that allowed them to pursue their higher calling. It was continued technological change that would free an increasingly larger proportion of the society from the necessity of physically debilitating work. The importance for scientific inquiry of enhancing the respect for the crafts, such as those who worked the furnaces, cannot be overstated. Whatever the particular medium, technology gives form to a part of human awareness that does not exist in nature or at least is not part of the direct unaided human experience of it. The "entirely novel phenomenon" that technology creates, "from molten metal to fired clay," often requires new "technology specific principles" (Schiffer 1992, 47).

This elitism, or artificial cleavage between thinkers and doers (who allegedly are not thinkers), has been widespread in human cultures and still can be found in modern Western society (Burkhardt 1952). However, in the European Renaissance, things began to change. There was an "erosion of the barrier between the scholar and the craftsman," which was a significant factor in the ensuing scientific revolution (Berman 1981, 57). The goldsmiths and other artist/craftsmen were perceived differently in the Renaissance. They gained respectability and were united with larger enterprises of the intellect. The artificial separation of mind and body was breached to the benefit of all human inquiry.

It is often argued that the Greeks and later the Romans stood on the threshold of the modern world. As Farrington (1944) points out, they stood on that threshold for several centuries and therefore "demonstrated conclusively their inability to cross it" (303). When the Renaissance and, later, scientific and industrial revolutions crossed that threshold, it was in part, at least in the opinion of J. D. Bernal (1971), because they broke down the social barriers of "ingrained intellectual snobbery that had barred the learned from access to the enormous wealth of practical knowledge that was locked in the traditions of almost illiterate craftsmen" (267). Chattopadhyaya (1986) makes a comparable observation for India on the social barriers to change resulting from the snobbery and elitism that separated workers from the world of knowledge (41). Ferguson (1992) speaks of the "historical significance of workers' knowledge" in fashioning and sustaining "the technological lead that Great Britain had over the Continent during the Industrial Revolution" (58). He adds, "The tacit knowledge and skills of workers may not have been the determining factors in Britain's leading role in the Industrial Revolution, but they were essential components of it" (59).

The importance of the workers' "tacit knowledge" did not end with the Industrial Revolution but continues to our present use of technology. Today, says Ferguson, "the knowledge and skills of workers—sensual nonverbal knowledge and subtle acts of judgment—are crucial to successful industrial production" (59). He believes engineers can learn "important lessons about latent possibilities and limits of craft knowledge and skills if they will but watch experienced workers in their expert, unselfconscious performances. And ask them questions" (59).

Many scientists and philosophers recognize the importance of tools (and machines) and experimentation in advancing science and knowledge of all kinds. To one philosopher, "science arises out of great leaps of practical ingenuity that enable scientists to acquire new data." There are practical struggles that determine whose machine gets built or whose theory gets tested. Once decided, says Galison (1997),[1] the "experiment goes forward. The result is determined by tools and nature, not by the ideology of the experimenter. . . . Like theoretical concepts, these knowledge-producing machines acquire meaning through their use . . . and in complex ways through their material links" to machines in other areas of inquiry (51). Dyson (1997) aptly speaks of "tool-driven" revolutions in science and argues that in some fields, such as biology and astronomy, the preponderance of scientific revolutions has been tool driven. "The effect of a tool-driven revolution is to discover things that have to be explained" (50–51). He explains, "If the tools are bad, nature's voice is muffled. If the tools are good, nature will give a clear answer to a given question" (1999, 34).

Romantics and Reductionism

To romantics, science is scorned, along with technology. Frank Kermode (1985) says the romantic writer William Blake "cursed Newton for ruining England by destroying the imagination." Yeats had similar criticism of Huxley and Darwin. Kermode (1985) adds, "to rich minds, magic, mystery, pseudo science, and apocalypse afforded considerable satisfaction." The real decadence was the "work of the men who imposed on the world what Blake called 'single vision'" (93). In contemporary academic parlance, "reductionist" science has been blamed for every sin imaginable.

Everyone accepts the whole to be greater than the sum of its parts, including modern scientists labeled "reductionists." What is ignored by the critics is that the whole, to be worthwhile, cannot be assembled from defective parts. Good scientists seek to get the parts right in order to construct the whole, and they look to the most heuristic theory (the whole) to guide them in their particular research in interactive process. The charge of reductionism is simply false, says Dawkins (1985). In fact, from the perspective of the layman looking into a field of inquiry through quality popular sources, scientists seem to be interested primarily in grand theories such as the cosmologist's quest for the "theory of everything." In truth, scientists are both reductionists and synthesizers, as the attempted construction of grand theories are based upon a vast assemblage of research rarely comprehensible to the

nonspecialist. For the nonscientist or for the scientist who has not succeeded at peer-reviewed research, the false criticism of reductionism is
a simplistic way of substituting an ideological shibboleth for a knowl-
edgeable, substantive analysis that requires technical competence and
an understanding of the issues involved.

A central theme of the philosophy of John Dewey (1934) was that
of breaking down the false dualism between thought and action and
reestablishing unity and continuity to the human life process. The task,
he says, "is to restore continuity between the refined and intensified
forms of experience that are works of art and the everyday events, do-
ings, and sufferings that are universally recognized to constitute experience" (3).

What Dewey and subsequent thinkers have been trying to do is re-
construct a unity of experience that has been inherent in the human endeavor ever since we became *Homo sapiens.* Dewey argued that artificial divisions of experience, the separation of practice from insight and
imagination, are not a product of human nature but of "institutional
conditions" (21). The role of thought is active, not passive. Mind is
"active and eager" and "lies in wait and engages whatever comes its
way so as to absorb it into its own being" (264).

This active dynamic human mind is facilitated by language, which
is a dynamic, open-ended process in the sense that we can continuously use it to create new combinations of words and ideas. Any sentence can be a combination of words and ideas that has never been previously expressed. What is almost miraculous about this process is that
communicating can take place not only in the present, but also in the
future with unknown others via writing and recording. In contrast to
language, signs and signals are static, limited in what can be commu-
nicated and incapable of generating novelty. In my analytical framework, the relationship of language to signs and signals compares to the
relationship of technology to tools. An individual tool, whether created
by humans or other animals, expands the resource base of its user but
is inherently limited and implies a finite set of resources. Other animals
have tools, but only humans and chimpanzees have "tool-kits" and
"tool-sets." Tool-kits exist when the same tool is used to solve different problems or different tools are used to solve the same problem. A
tool-set exists when "two or more tools are used sequentially to
achieve a single goal" (McGrew 1993, 158–59). Only humans have
tools used to make other tools and have the kind of foresight and planning to transport raw materials, thus creating workplaces and tool

caches (McClellan and Dorn 1999, 8). This same foresight made tools an integral part of complex hunting strategies (Mithen 1996, 96, 104, 167). Only humans keep tools, rather than discard them. Tool-sets and tool-kits are important differentiating characteristics; taken alone, they do not constitute technology.

Insofar as a tool is part of a larger technological process embedded in a web of ideas, there is a creative combinational potential in which technological possibilities and resource capacities are open-ended and limited only by human imagination or the lack of it. This larger process of technology involves specialization, social organization, and a social process of learning and knowledge transmission. "It is precisely this type of socially organized and cooperative division of specialized labor that is *never* seen except in human tool manufacture and use" (Wilson 1998, 171). Wilson adds, "No matter how clever chimps may be—or how much they watch and learn or interact with one another—they do not cooperate in the construction and use of tools" (171–72).

In my scheme, technology is first and foremost ideas but does not exclude hands and the material manifestation of a tool or technology. The hand has been the primary pathway for the material embodiment of these ideas as tools. Pinker (1997) says, "Hands are the levers of influence on the world that make intelligence worth having" (194). The hand has been the mind's instrument for the creation of the plastic arts and for the performance of music. The interactive processes of tool using and the evolution of protohominids to humans involved concomitant transformations in the hand and the areas of the brain that control the hands. Sherwood Washburn (1960) has described an interactive process by which those protohominids with a larger area of the brain controlling the hand and improving the ability to create tools had a survival advantage. Reciprocally, improved tools gave a survival edge to those with a larger area of the brain controlling the hand in an ongoing evolutionary process of technological and human evolution (DeGregori 1985, 12). Though the changes in the hand and the brain may have been random, tool using defined the selection and survival mechanism, giving a direction to the process that is nonteleological. In many respects the hand evolved as a somatic instrument that creates and uses the extrasomatic instruments that were the joint product of the hand, mind, other extrasomatic instruments, and social organizations working together.

Bernard Campbell (1982) calls the hand a "marvelous tool" and argues that it is "used to full value only when it manipulates other tools"

(47). Today we can understand the relationship between the hand and the brain better than those who have expressed disdain for people who worked with their hands. In fact, the hand is a magnificent instrument. It has twenty-five joints and fifty-eight distinctly different motions. Campbell asks us to "imagine a single tool that can meet the demands of tasks as varied as gripping a tool, playing a violin, wringing out a towel, holding a pencil, gesturing, and—sometimes we forget—simply feeling" (47).

Implicit in all these actions of the hand is the fact that the hand is a vital mechanism by which humans learn about the world. The brain sends messages to the hand to cause it to act. The purpose for these "messages going to the hand in the first place" was so that the hands could "reach for, grasp, touch, turn, weigh, join, separate, bounce, and so on, whatever it was that came into their possession." In other words, the hands were obtaining "information that could be obtained *only* by acting on the object being held" (Wilson 1998, 275–76). Succinctly stated, "the hand speaks to the brain as surely as the brain speaks to the hand" (Davies 1985, quoted in Wilson 1998, 60, 276).

The idea that we learn through our hands has antecedents but the attribution was generally not complimentary until recent times. Until very recently, writing on the American Indian reflected these biases. At the beginning of the twentieth century, W. J. McGee argued that the hand as a learning instrument was a mark of primitivism. He said, "experience has shown that among all the aboriginal peoples, the hand leads to the mind" (Dilworth 1996, 151). At about the same time, a distinguished anthropologist writing on the Indians of the American Southwest established a hierarchical taxonomy in which the mental stage of human development was the highest (Cushing 1892, 289–91). Cushing "postulated three stages of intellectual development: the biotic, the manual, and the mental. The biotic was the earliest stage, where humans were barely human but had developed hands, which they used in climbing and fending and defending. The next stage, the manual, was when humans began to develop 'extranaturally,' when they acted upon (made) the environment" (Dilworth 1996, 151–52).

In Cushing's hierarchical schema, the way in which humans use their hands "frames the mind" and the way in which humans manipulated the environment was a "determining factor in human evolution" (Dilworth 1996, 152; Cushing 1892, 291). Cushing was definitely not alone in his thinking on the importance of the hand in primitive culture. Tourist literature at the time referred to the arts of American Indians as being all "done from memory and with their hands" (Dilworth 1996, 155). Dilworth summarizes the thinking of the time on the hand and the mind as follows: "The centrality of hand labor in characterizing the primitive mind was the subject of much scientific discourse." Dilworth adds that since American Indians were seen as survivals of the "manual stage of cultural evolution," an opportunity was provided for scientists to study these artisans "to discover conditions of invention and creativity" (151, 154).

There is an interesting duality operating here. Though the civilization based on the mind was considered by many to be superior to that of the hand, nevertheless, acquisition of the products of the hand carried and continues to carry a certain cachet and prestige for those who are able to afford them. In addition, contemporary New Agers and others have inverted the hierarchy and have found a special virtue in preindustrial cultures based on manual labor. This has been particularly true of the New Age literature on the American Indian.

The cleavage between the thinkers and the doers that I have criticized is in fact artificial precisely because it denies and in some respects prevents that basic process of learning that was an integral part of the evolutionary developments that made us human. If, as Washburn contends, it was the interaction between tools, hands, and brain that brought us to our status as humans, why should we abandon it? To believe that the brain is the sole basis of intelligence or the sole source of knowledge is what Wilson (1998) calls the "cephalocentric view of intelligence" (295).

Ingold (1993) challenges the dichotomy that "gives us a distinction between intellect (as a product of mind) and behavior" (as bodily execution). He disputes the idea that intelligence is the "operation of a cognitive device . . . which is somehow inside the animal and which, from this privileged site, processes the data of perception and pulls the strings of action." Rather, the argument is made that "like locomotion, cognition is an accomplishment of the *whole animal,* it is not an accomplishment by a mechanism interior to the animal and for which it serves as a vehicle" (431; Wilson 1998, 295–96).

Wilson (1998) is critical of the "scientific reductionism" that reduces human intelligence to one generalizable phenomenon. He believes we have many different specialized intellectual capabilities so that it is more sensible to speak of "intelligences" (305, 307). Here we are taking the first steps away from focusing on a hierarchy of mind and body or of elitist versus lesser uses of the mind toward more operational concepts as to how mind and body work together, how we individually and collectively came to have the rich and diverse capabilities that we now have. Most important, we must continue to explore how to exploit this knowledge to create the environments that allow us individually and collectively to use the means at hand for the maximum individual and collective development of our intelligence.

It is increasingly recognized that the human brain, the hand, language, and other elements of the "human behavioral evolved together as one complex adaptive and cognitive whole" (Gibson 1993a, 9). Similarly, Pinker (1997) says, "precision hands and precision intelligence coevolved in the human lineage, and the fossil record shows that hands led the way" (194). Wilson refers to "evolution's remodeling of the hand," which opened the "door to an enormously augmented range of movements and the possibility of an unprecedented extension of manual activities." The evolution of the hand did not occur in isolation. Wilson observes, "As a collateral event, the brain was laying the foundations of cognitive and communicative capacity" (58). Anatomically, as would be expected, the arm and shoulder were evolving along with the hand and the brain to create the complex structure that allows the hand to perform the functions unique to humans (Wilson 1998, 58; Pinker 1997, 11).

There is a growing body of literature that builds on Washburn's insight relating the development of the brain to the evolutionary process of the brain's interaction to the evolution of tools. Truly outstanding books like *The Hand* by Frank Wilson and *Tools, Language, and Cognition in Human Evolution,* edited by Kathleen Gibson and Tim Ingold (and the many books and articles that they cite), are but a sample of the fine work that is being done relating tool use to the brain developments that facilitated the growth of language and cognitive skills.

The ultimate irony of snobbery that belittles working with one's hands, tool using, and technology is that the so-called higher functions of the human brain and thinking are themselves a product of tool using, as are the brain's nonmaterial instruments, language and cognition. Deacon (1997) calls the interaction between language and evolution the "co-evolutionary net" (349–65). "The evolutionary dynamic between social and biological processes was the architect of the human brain" (349). Deacon recognizes the analogy between the way language developed for one application and was then used for many other purposes, and the manner in which other technologies had original uses that gave rise to a huge number of unexpected uses (350). He even says that the forms of behavior that emerged from early evolutionary developments played a critical role in the dynamic that made us human. "The first use of symbolic reference by some distant ancestors . . . has affected hominid brain evolution ever since." Just as in my schema, tools are the embodiment of ideas, "the physical changes that made us human are the incarnations, so to speak, of the process of using words" (350).

Aime Cesaire, the Martiniquan poet of African extraction, may have, in his poetry, sung "hurray for the those who never invented anything, hurray for those who never explored anything," but most leaders and thinkers around the world seek the benefits of technology for their people. The great twentieth-century Chinese philosopher Hu Shih (1928) long ago questioned the validity of the alleged "materialism" of the West as contrasted with the "spirituality" of the East. He found little spirituality in the condition of the rickshaw coolies, "those poor beasts of burden who run and toil and sweat under that peculiar bondage of slavery which knows neither the minimum wage nor any limit of working hours" (27). Hu Shih considered the terms *materialism* and *spirituality* to be misused. To him, the term *material civilization* should have a neutral meaning because "all tools of civilization are the material embodiments of ideas," which is as true of the earliest stone tools as it is of the most complex modern technologies. Nor should the designation "materialist civilization" be considered a stigma or necessarily applied to the technologically more advanced peoples (27).

In contemporary Asia, the term *Asian values* has a complex set of meanings, but one common dimension and use of that term is the argument that these values allow the more effective implementation, integration, and utilization of advancing technology. Any misplaced beliefs that modern technology and science were an exclusive preserve of Western culture, or that Asian or other non-European cultures were incapable of scientific, technological, and economic advancement, have been massively dispelled by the economic performance of the Asian countries over the last few decades. Hu Shih correctly recognized that the civilization that understands the potential of technology "will continue to grow and benefit itself" (27). He continues, "to me, that civilization is materialistic which is limited by matter and incapable of transcending it; which feels itself powerless against its material environment and fails to make the full use of human intelligence for the conquest of nature and for the improvement of the conditions of man" (27).

Technology: The Humanizing Endeavor

The complex of technological, biological, and cognitive change that made us *Homo sapiens* also made us human. The qualities of thought and sensitivity to one's environment that are often defined as the highest aspiration of humanity were inherent in the tool-using, tool-combining, resource-creating process even before this transformation occurred. From earliest tool use, says Marshack (1972), there was the necessity for a hominid to have "a working knowledge of time, place, and the direction and bounds of his effective territory, with a working knowledge of materials" (115). Marshack describes in some detail the vast amount of knowledge, not only "of sky, season, and animals, but of materials and processes" that was part of the "store of skills and knowledge" of the Upper Paleolithic hunter (131). The flint knapper had to look at a rock and see a tool embedded in it, just as an artist sees a sculpture in a piece of stone or wood. The perception of potentiality and therefore the possibility of human transformation of the environment are essential components of the process of tool creating and using.

In my book *A Theory of Technology* (DeGregori 1985), I stress that technology is first and foremost ideas. The stone tool was an idea in the knapper's mind before it was a tool and could become a tool only if it was preceded by an idea. But equally important, it could not become a tool without the interaction of the brain and the hand transforming the material substance (Gowlett 1992). Not only does the idea precede the tool making, it continues as an integral part of the process until the tool is completed and ready to use. The use of tools is a continuation of the ideational process of their manufacture and a continuation of the interaction of the brain and the hand. Reader notes, "There was nothing arbitrary about the manufacturing process. The long axis, the cutting edge, the point and the symmetry of the carefully controlled curves were imposed on the stone, they did not arrive by chance" (79).

In every step in the tool-making process, the knapper was acting in terms of knowledge, experience, and ideas: "Each blow struck on stone opened up new possibilities for that. The process could be controlled only by constantly comparing the work in hand with an image of the finished product that is fixed 'in the mind's eye'" (Reader 1998 79).

Among the definitive characteristics of *Homo sapiens* and the emergence of the modern mind was an outpouring of creative art in rock paintings and in small statues. Humans "using visual media" and "conveying ideas so complicated as to totally confound us" were producing "their own world, a symbolic world that included imaginary creatures and deities." The story is one of creativity and accomplishment where "art and survival were tightly intertwined." Art and tool using became a universal characteristic of the human endeavor. Humans became capable of what Schiffer called "holistic problem solving" (Mithen 1996, 15, 155–63; White 1986, 104; White 1986, 35; Schiffer 1999, 64).

Human nature was no longer one of the qualities of a *being,* but was part of the potential of *becoming.* We could think in terms of time, which is both an abstract symbolic concept and a vital necessity for a predator that is slower than its prey (Marshack 1972, 113, 116, 370). Thinking in time involved thinking in process, as the stages of the development of a plant or the migration patterns of animals became forms of knowledge that were part of the technological knowledge for survival. With clay pottery, agriculture, and metals, thinking in time and the potential of becoming were transformational as humans became causal agents in a productive process. Calvin (1993) says that the same "neural circuitry" in the brain for language and tool use and even for throwing (as a spear) are also used for planning ahead. "For toolmaking and tool use, one usually has to make a novel *sequence* of movements" (231). Gibson (1993b) argues that for humans, "tool use, social behavior, language and mathematical thought are mutually interdependent and reinforcing." They are fundamental to the "human cognitive niche" (256). He adds, "Human technology . . . is inextricably linked with social structure. The human life style is not only tool-based, but many tools and constructions are used and reused over long periods of time. This long term use of the same tools and dwellings is a distinguishing feature of human adaptation" (256–57).

Individual and socially organized planning is an inherent characteristic of our species. Elsewhere, I have argued that beyond the most basic forms of hunting and gathering in tropical climates, there is no natural fertility of the soil or natural carrying capacity for humans; all have been created by human ingenuity or technology (DeGregori 1985; 1987a,b). We evolved in an interactive process with our tools and with the environment that we adapted to our needs. Thus, "tools are us" (Schick and Toth 1993, 52).

The action to create our future fits within Dewey's (1958) conception of the "artistic act" as being a "revelation of possibilities hitherto unrealized" (359). Culture becomes "designs for living" (Geertz 1977,

31). The quality of our designs determines the viability of our existence. "Tools, hunting, family organization, and later, art, religion, and a primitive form of 'science,' molded man somatically; they are therefore necessary not merely to his survival but to his existential realization" (Geertz 1973, 83; 1977, 30–31). Geertz (1977) adds, the "'slogan man makes himself' now comes to have a more literal meaning than originally supposed" (27).[2]

The intertwining of art and survival was not accidental. The creative processes of art and technology are of one piece. The cognitive achievements implied by art were part and parcel of the conceptualization and operation of societal structures that organized economic and technological activities. The intimate and vital interconnection between technology, time factoring, and the creative act meant that humans, in determining their future, could begin to contemplate alternative futures. We became a creature possessed by the necessity of making long-term choices. Given knowledge of these choices, as the existentialist philosophers tell us, not to choose is itself a choice. And normally it is a bad choice, since it denies over thirty thousand years of the conditions that made us distinctly human.

The cognitive revolution and time factoring made choices as to which technological and demographic pathways were possible. Early in any process, empirically based knowledge is limited and, therefore, so is the range of knowable choices. Consequently, some of the most important choices may have been fortuitous. As Hayden (1981) notes, "domestication did not first occur where the environments were rich enough to support sedentary, hunting-gathering based, ranked societies" (530). One factor in domestication may have been an environment where limits of the existing technology, namely "resource stress," have occurred. Even so, Hayden believes that there was the possible choice of attempting to balance morbidity/mortality or of minimizing "the effects of resource stress" by innovation and transformation. "Only man, however, had the technological potential for making choices in accordance with the second" (523).

Drury (1998) argues that humans can occupy almost any habitat and move freely among habitats. We adapt to new environments, and we also adapt habitats to our needs. And we have made an additional critical leap. Humans do not need to wait for slow, undirected variation and Darwinian selection. Instead, we identify problems, develop solutions to them, and pass the solutions on to others. Lamarckian evolution by the inheritance of acquired characteristics may not be true for biological evolution, but it is for cultural evolution, thereby increasing the rate of change. As humans, "we are not a specially created miracle, nor are we an aberration, but we are very, very special" (183).

Technology, Alienation, and the Life Process

Hu Shih (1928, 26) notes, "Man is a tool-making animal and it is tool making which constitutes civilization." Technology is often seen as simply an extension of human capability. Culture, including technology, has been similarly defined. Geertz (1977) counters that "rather than culture acting only to supplement, develop, and extend organically based capacities genetically prior to it, it would seem to be an ingredient to those capacities themselves" (28). He goes on to indicate how meaningless it is to conceive of a "cultureless human." Included in the analysis is the *Homo sapiens'* brain, which "having arisen within the framework of human culture, would not be viable outside it" (Geertz 1973, 68; 1977, 28). We are incomplete or unfinished animals who complete or finish ourselves through culture (Geertz 1973, 49). Rabindranath Tagore earlier expressed similar sentiments in saying "man has a feeling that he is represented in something which exceeds himself. He is aware that he is not imperfect but incomplete. He knows that in himself some meaning has yet to be realized" (Ghosh 1987, 153).

There is at least one aspect in which technology extends our senses. Technology is also much more than an extension. Our intellectual grasp far exceeds our biological reach. With telescopes and microscopes we can see farther and smaller. A myriad of technologies used in archaeology, paleontology, and geology allows us to study the history of human life, of life on earth, and of earth itself. In our contemporary world, humans can use a variety of other technologies to explore the depths of the oceans, the tops of mountains, or the upper canopy of the rain forest. These and many other technological extensions of our senses are equally extensions of our minds as we can explore and understand the world around us and the world beyond us from the smallest particles to the distant cosmos. In other words, we are tuning into nature. All of these involve areas of scientific inquiry that also have been made accessible to the layman through popular books, on the internet and through film and video (Lucky 2000). We are on the verge of having the "virtual observatory" using the internet to link the world's great observatories, making their digital observations accessible to anyone on line (Schilling 2000). Each of these could be used to illustrate my point that our technology connects us to nature in numerous exciting and informative ways.

Romantics do not seem to realize that all technological and scientific inquiries are simply different ways of accessing and understanding the world (nature) around us. To those poets who claim that science somehow takes away from the beauty of the stars, the physicist Richard Feynman counters that he too sees the stars and feels them, allowing the "vastness of the heavens" to stretch his imagination (Baeyer 2000, 14). Speaking of that vast pattern of which he is a part, Feynman adds: "It does not do harm to the mystery to know a little more about it. For far more marvelous is the truth than any artists of the past imagined" (Baeyer 2000, 14).

Feynman turns the tables and asks why poets cannot find beauty in scientific truths, such as physics and astronomy: "Why do the poets of the present not speak of it?" (Baeyer 2000, 14). Feynman further asks why humans can write poetry about Jupiter as a man or a god but they remain silent about Jupiter as "an immense sphere of methane and ammonia" (Baeyer 2000, 14).

In an earlier book, I commented that though the immensity of the universe in duration and in size, from the subatomic particle to the billions and billions of galaxies, may make us feel small, "it does not make humans feel insignificant" (DeGregori 1985, 84). It is the human mind using its creation, technology, which has explored and made this immensity known to us. If we are somehow compelled to feel humble before all this, we should also feel humble before those among us who have so expanded our horizons. Their depth of understanding yields so much for us to know and appreciate and shows us that there is so much more that we wish we could know.

With the education and leisure modern society allows us, one wonders how many amateur geologists, archaeologists, paleontologists, astronomers, and botanists there are who are expanding their knowledge and ours. For those not actually going out into the field, there are planetariums, natural history museums, and nature films that reach millions of people. It may be argued that none of these are adequate substitutes for the "real thing." However, the "real thing" is not a substitute for what technology brings us. They are different and complement one another. Those who look up to the stars each night see only a small portion of the sky and only as it exists at that moment. Planetariums show many different areas of the sky and are able to create very real simulations of the heavens as they were or will be. The same can be said for films or simulations in other areas. Today, we can see recreations of the past and projections of the future, and we see nature as it now exists throughout our planet. Experienced photographers and filmmakers can capture events within plants or animal burrows or on the tops of trees in rain forests that would not be accessible to us even if we were there. These amaze and entertain us but they also richly inform us.

Alienation

It is ironic that the technology that has allowed us to achieve so much is considered by many to be alienating, thereby making us spectators to experience, but not participants in it. The term *alienation* refers literally to separation and is used in respect to the individual and his or her environment. In popular discourse, it is frequently treated as an absolute—one is or is not—rather than a condition of more or less. Yet some degree of alienation is essential for the human life process. A membrane alienates the cell, allowing its internal components to interact and the cell to interact with other life forms and the environment. Without some degree of alienation, there is no individualization, no communication, and no development of complex forms. Without an ability to build complexity, "there was no indication that the earth's patina of life would ever be more than inconspicuous scum lying on the ground" (Margulis and Sagan 1986, 113).

The nature of civilization involves some degree of alienation from nature. Agriculture, a pillar of all civilizations, requires alienation of some plant life forms. Domestication of plants and animals gave rise to life forms that could not continue to exist without humans providing some separation from nature. Plants and animals are protected from predators and competitors. Agriculture has provided ever-greater buffers from the vagaries of nature. For example, irrigation provides water when rain is inadequate or fails completely. These various forms of alienation have provided a regularization of food supply, which has saved tens of millions of lives in our time. In this respect, alienation is good.

The process by which we made the transition to *Homo sapiens* involved art and alienation. Though physiologically the transition to *Homo sapiens* may have taken place as early as ninety thousand years ago, that later (about thirty-five thousand years ago) complex of activities by which the cultural aspects of this transition are defined—namely, fire, ritual, and speech—included art as an integral component. Drawing on the ideas of Schiller (1954), John Halverson (1987) says, "Man began as a sensuous animal whose relation to nature was simply one of response: nature encountered rather than perceived. Nature existed only as it existed for man" (67).

Fortunately, humans did not remain at this level limited to a pure, direct experience of nature. Man became man, says Halverson, when man was able to see natural appearances as things in themselves. The subject-object relationship was born at this point, and nature became an object rather than merely a force.

Novelist Edward Abbey (1968), describing time spent outdoors, stated that when he walked at night in nature, he kept his flashlight off so that his eyes became accustomed to perceiving the broad expanse and nuances of reality rather than that which was in the narrow beam of the flashlight (14). In reality, his observation carries a double edge: We can turn the flashlight off and on. Our technology allows us that concentrated, analytical focus on otherwise nonobservable aspects of reality while we temporarily alienate ourselves from larger realities. No one would argue for total alienation, total dependence on, or uninterrupted utilization of a technological means of analysis. Intelligence allows us to use both in whatever proportions best suit our purposes. Specialization and division of labor allow us to use the narrow and the broad observations of others. And intelligence and memory allow us to integrate these into a meaningful comprehensive, unified experience. The core of being human is to do not one or the other, but both.

The "architecture of the human mind" is such that humans can react rapidly, says Fodor (1983), almost instinctively in some circumstances such as sudden danger, and in a slow, reflective manner, integrating many different types and sources of information in other circumstances. Both are essential to our survival as a species, and fortunately, we have them both and can choose to use one or the other as warranted (Mithen 1996, 39). In Mithen's (1996) words, we have a "flexible mind" characterized by "cognitive fluidity" (185, 195). We evolved these diverse mental capabilities for a reason. We can fault someone for reacting when they should be contemplating or vice versa, but we can not fault human beings for using all their capabilities at some time or other. Abbey and others who might wish to exalt one mode of thought to the exclusion of others are restricting the human potential rather than expanding it, as they seem to believe that they are doing.

Clothing and shelter have provided that measure of alienation from harsh climates, allowing tropical species such as us to survive in a variety of habitats. Without technology, we would be a species limited to the tropics and always under the threat of extinction due to a climatic or other environmental change. Humans have used technology to adapt the environment to their needs rather than the slower, less certain process of biological evolution to adapt the organism to the environment. Consequently, humans are the only mammals to have colonized the entire globe without speciation. Had we gone the unlikely route of speciation sufficient to adapt to all climates, we would still be far from covering the planet. Given that no other primate group has been able to inhabit the coldest climates, it is not likely that our lineage would have done so. Even had we done so, there would be different species and no "we" to experience "nature" in all its glorious diversity, sharing the experience with others. In a word, technology as a form of alienation has allowed humans to experience forms of nature in various places and climates that would have been otherwise not accessible on a sustainable basis.

Something Lost?

The view of technology alienating humans from the environment is deeply ingrained, as is the vision that we have lost a natural condition that made us one with nature. The poet Robinson Jeffers (1941), a favorite of modern antitechnologists, spoke of "man disseevered from the earth and stars and his history." He spoke of "organic wholeness, the wholeness of life and things, the divine beauty of the universe, love that, not man apart" (594).

To many, the organic unity of society that we once enjoyed—both among ourselves and with nature—has been lost. To some, it is a loss that has occurred within modern memory. Yet if we go to the periods of the recent past about which some are nostalgic, we find that they too were bemoaning the loss of a sense of shared community. The question as to how we can have "organic wholeness, the wholeness of life and things" without including our technology and science is not answered because it is never asked. As Levenson (1994) reasons,

> Remembering that science is a form of art enables us to see that it is not simply—or even so much—about nature out there as it is about ourselves, about making sense of the human condition. . . . Science, even the most abstract of inquiries, remains absolutely a human passion (316).

Something Found

It has been technology that has allowed those whose senses are debilitated to interact with others and the cosmos. Throughout the world, there are physically challenged individuals who are learning, creating, and moving about thanks to a variety of technologies. Two extreme and inspiring examples are Christopher Nolan, winner of a coveted literary prize for *Under the Eye of the Clock,* and Stephen Hawking, one of the world's greatest physicists and cosmologists (Nolan 1988; Hawking 1988). One is exploring the limits of human emotion and the other is exploring the limits of the cosmos in space and time. Neither could function nor communicate without technology. With technology, however, they and others have broadened horizons and deepened the understanding of us all.

In many respects, we are all bound by our senses. We may not have the courage of Nolan or Hawking, yet we are all prisoners of our bodies, limited in what we can experience through our senses. With technology, we overcome these limits and reach out to encompass the stars within the sphere of our understanding. The "organic wholeness" that Jeffers proclaims and others echo is continually being achieved and extended with technology.

There is plenty of discovery and excitement in science and technology to satisfy even the most jaded among us. As Maddox (1998) cogently argues, even more exciting than finding answers to questions now "crying for our attention" is the excitement that will come from finding answers to questions that we do not yet know enough to ask (1). Are there things that people were not meant to know, as the story of Frankenstein tells us? Ask the disabled whose lives have been improved (or enabled) by technologies that are continually being created and improved. Ask those who have had their sight restored or improved with bioengineered eye tissue if they are opposed to biotechnology (Tsai, Lia, and Chen 2000; Schwab and Isseroff 2000a,b; Dobelle 2000).

Symbolic Thinking and Human Development

Symbolic representation of any kind, in language or in art, involves an abstraction from reality and therefore some intentional alienation. In many respects, the development of the brain involves alienation for animal life. Experience is filtered and selected physiologically and

mentally so that the organism can respond to the stimuli—food or danger—that allow it to survive. As with the smaller units of life, for humans some alienation is necessary for communication. Having acquired a degree of detachment or alienation, humans have developed intelligence, knowledge, and technology, which, as we have noted, more than compensate for what we have lost. Even the development of the physical mechanisms for human speech involved biological losses in the human ability to take in oxygen, but this loss became a gain with enhanced knowledge and communication (Lieberman 1988, 23–29; Crystal 1987, 291).

Throughout the history of Western technology and science, ideas and practices that furthered the expansion of knowledge and the arts were greeted with suspicion. Moreover, they were vigorously opposed. In hindsight, it is hard to understand why early forms of writing or new styles in painting or printing would be opposed, since we can see the enormous benefits to human well-being that flowed from them. Similarly, with new technologies that I will discuss later, the benefits have been so great that opposition to them is not always understandable. Still, there is the fear that new technology will lead to something important being lost.

The Alphabet

Our world today is unthinkable, even to the most mossback reactionary, without the alphabet and the writing system it permitted. Yet, this technology was initially opposed.

The development of the Greek alphabet, says Havelock (1982), was "a piece of explosive technology" (6). The significance of the Greek alphabet was that the letters were "meaningless"—they were divorced or alienated from the reality that they were to be used to describe.

> A true alphabet, the sole basis of future literacy, could only become operative when its components were robbed of any independent meaning whatever in order to become convertible into a mechanical mnemonic device (84–85).

Because the alphabet was meaningless, it could be used for more than one language and was the technological secret that accelerated "the process of cross-translation between them" (85). Discourse became conceptual, artifacts and ideas were preservable without recourse to memory, and it released "mental energies . . . contributing to an im-

mense expansion of knowledge available to the human mind" (9, 87). Philosophical ideas such as "atomism and the alphabet alike were theoretic constructs, manifestations of a capacity for abstract analysis, an ability to translate objects of perception into mental energies" (82).

Another important attribute of the Greek alphabet was that it democratized literacy or made democratization possible. It removed the "elitist status of literacy characteristic of craft-literate epochs" (83). Everyone could partake, not just a few. Democratization tends also to be decentralizing and participatory, qualities that are generally not identified with technology by its critics. The alphabet embodied potential for widespread literacy, but it was not manifested in the world of its Greek inventors (Harris 1989, 14–15, 94–96).

Writing

Plato's ideas have been transmitted to us by writing, yet ironically he was a severe critic of it. According to Ong (1982), Plato thought of writing as "an external, alien technology, as many people today think of the computer" (81). Ong goes on to say that "writing (and especially alphabetic writing) is a technology, calling for the use of tools and other equipment: stylus or brushes or pens, carefully prepared surfaces such as paper, animal skins, strips of wood, as well as inks or paints, and much more." In the *Phaedrus,* Plato tells us (as always, speaking through Socrates) that writing "will produce forgetfulness in the souls of those who have learned it." Students will no longer "need to exercise their memories, being able to rely on what is written" and, horror of horrors, they will lose "wisdom" because pupils "will be widely read without benefit of a teacher's instruction" (Plato 1956, 68).

The idea that we will lose "something" because of a new technology has been with us since Plato and probably before him. No wonder modern cultural authoritarians of different persuasions would have us return to thoughts of Plato. However, their advocacy is most often in written form. Following Havelock, Ong (1982) also finds the Greek alphabet to be "democratizing in the sense that it was easy for everyone to learn" (90).

Many wish to compromise with the critics of technology by proclaiming the neutrality of technological change. However, I understand technology as the capability of doing something that is desired, and progress in technology is an enhanced capability to achieve a desired objective. If one defines language as a technology, then all art forms involve the use of some technology (except possibly the art of mime).

Those forms that initially use only one technology, such as language for storytelling, will through time have their artistic potential expanded by new technologies. The continued growth of technology and its use in an artistic endeavor will often lead to the creation of new art forms. Drama is one of many art forms that emerged out of earlier forms, mid-wifed by technology.

Writing as a technology used various instruments. Later, printing provided a greatly improved mechanism for artistic experience to be expressed, accumulated, and stored. This function of printing for the arts was no different from the role it played for the creation, cumulation, and storing of scientific knowledge or knowledge in other areas in the endeavor of being human. The process of cumulating knowledge and experience has always been a stimulus to continued creativity in the arts and sciences. As author Jorge Luis Borges proclaimed, the book extends not only our memory, but also our imagination (Rybczynski 1983, 4).

Printing

Few would argue against the importance of the fifteenth-century advances in the arts and technology of printing to the Renaissance in Europe. Elizabeth Eisenstein (1983, 432–33) notes, as have others before her, the stimulus to creativity and the combinational possibilities for individual scholars or scientists of having many more works of others available to them. It is a principle widely recognized in other areas. In economics, since Adam Smith, extending the market gave an increased potential for division of labor, specialization, and economic advancement. The current push for globalization is a further recognition of this principle. Division of labor, specialization, and extending the market, including globalization, all require advances in technology.

For scientific activity, Eisenstein (1983) argues, the technology of printing, while facilitating the interchange of information by means of greater access to the "little books of men," advances our knowledge of the "great book of nature" (186). The "slavish copying" necessitated by the lack of a mechanical means of reproduction prior to printing was unlikely to give rise to the critical judgment that became the hall-mark of the Renaissance and the scientific revolution that followed. Printing clearly contributed to the critical attitude that produced "a distrust of received opinion and a fresh look at evidence" (2). Science as we know it is unthinkable without free and open inquiry.

Antitechnology and Medical Care

New technologies such as writing or printing often bring a fear that "deskilling" will result. At the beginning of the twentieth century, many physicians were fearful that the blood pressure cuff would "pauperize" their "senses and weaken clinical acuity." Another physician complained that it "would deal a death blow to the painstaking study of the pulse" (Cowan 1997, 202). Today, few would argue that the flood of diagnostic technologies has deskilled the medical profession. In fact, taking blood pressure has become routine, using a technology that is viewed as basic, not high tech. The blood pressure cuff and the simple and much more complex diagnostic technologies that followed have required physicians to expand their skill capability to utilize the information provided by the new technology.

The Instrumentation Dilemma

Nearly a century before the blood pressure cuff, the invention of the stethoscope brought a similar response. The stethoscope greatly enhanced and supplemented the ear in allowing for what was called "mediate auscultation," or the use of an instrument for diagnostic monitoring of sounds made by the body's internal organs, listening for abnormalities. Some said the stethoscope allowed the doctor to "see" the disease by listening with the "ear becoming an eye through auscultation" (Reiser 1978, 30). It allowed diagnostically significant access into the living human body without being invasive. In spite of the "clear superiority of auscultation in diagnosing chest disorders," the older ways of medical practice were still supported by many. Physicians "were reluctant to abandon techniques that had cost them so many hours to learn," or were unable to command the new technology—or thought they could not (32). Another "deterrent" to the use of the stethoscope was that it "required the use of the hands and an instrument," which might cause physicians to be classed along with surgeons, "as mere craftsmen" (37). Surgeons were the "instrumental" branch of medical science, which was a term of reprobation and not praise (38).

An antitechnological bias in medicine had its origins in the medieval university where medical training took place, along with that of theologians and lawyers. Physicians accepted the beliefs of theologians and lawyers that "the use of tools and manipulation lowered the esteem of and social standing of scholars and physicians" (Reiser 1984, 7). In the medieval university, learning was book learning. In practicing their profession, "scholars could not act like tradesmen, with whom the use of instruments was associated." This "prejudice caused surgery to become disconnected from medicine," as surgeons were compelled "to set up their own schools outside the university." Physicians who "established manual contact with patients" were considered to be lowering themselves and were "derisively labeled 'body physicians'" (7). Both surgery and medicine were harmed by the separation. Not until well into the nineteenth century, with the overwhelming diagnostic success of a series of basic medical instruments, beginning with the stethoscope, was the "surgical approach reintegrated" into medical practice (7).

The zeal of the early proponents of the stethoscope and subsequent medical instruments led many of them to disregard entirely the older techniques of observation, taking a patient's history, and inquiring of the patient as to what the aches and pains were and where they were located (Reiser 1978, 31–36; 1984, 7, 227–31). Without being fully aware of what was happening, the physician in the last two centuries has gradually relinquished his or her "unsatisfactory attachment to subjective evidence—what the patient says—only to substitute a devotion to technological evidence—what the machine says." In becoming overly dependent on instrumentation, the physician "exchanged one partial view of disease for another." For the physician, these circumstances tend to cause estrangement from the patient and from personal judgment (Reiser 1978, 230). No matter how sophisticated instrumentation is and may become, patient histories and perceptions remain a vital component of medical practice.

What Critics Say

Postman (1992), a perennial critic of technology, uses the Reiser analysis of medical instruments to argue that a "technological imperative" exists by which the physician is estranged from his or her own judgment. There is no longer a multiplicity of ways of "treating illness; there is only one method—the technological one" (101–2). In this line of thinking, doctors do not use technology but are used by it. Technology is not "neutral." It creates its own imperatives along with a "wide-ranging social system to reinforce its imperatives." The "reign of technology" is likened to a disease (105).

At the risk of sounding like someone who has yielded sovereignty over his life to technology, allow me to say that all of my past and present doctors keep a record of my medical history, sit down prior to each examination and ask questions and then record my answers with pen and ink on paper. While there may not be enough of this old fashioned but still valid component of medical care, is it the fault of the technology? Even Postman, who makes a valiant but largely unsuccessful attempt for "balance," has to concede the extraordinary gains wrought by medical technology.

Postman's assignment of blame to technology does not stand up to closer scrutiny. First he charges that doctors in the United States use far more technology and do far more invasive procedures than their European counterparts, yet our life expectancy is no greater than Europeans'. Even if there were no other explanation for these discrepancies, it does not provide proof for a "technological imperative" unless Postman can explain why European medical practitioners are less susceptible to it, particularly since an "imperative" presumably leaves us no choice (Postman 1992, 94–95).

Postman unwittingly suggests an institutional explanation when he asserts that if a physician in the United States does not use all the "available technological resources, including drugs," he or she could be deemed incompetent or even subject to a liability lawsuit if something goes wrong. Equally important is that in our third-party-payer system, "doctors are reimbursed by medical-insurance agencies on the basis of what they *do,* not on the amount of time they spend with their patients" (Postman 1992, 102). Without question, this latter point is an institutional critique, not a technological one.

The institutional issues of modern medical care are extremely complex. The benefits from the care are obvious, but the cost of providing them is great, such that, except for the very rich, anyone without medical insurance is considered to be at risk and severely disadvantaged. How to make the advantages of medical care available to all is a serious problem and one that Postman in his antitechnology zeal neglects to mention. And how do we make physicians responsible and legally liable without their overtesting and overusing technology (and other defensive practices), which greatly raise the cost of medical care?

To keep costs of medical insurance affordable, there are various schemes of "managed care" that control the patient's access to diagnostic technology and intervene between the physician and the patient. No system, however well intentioned and well designed, is perfect.

Rules to limit access to medical technology and prevent costly overuse
inevitably will deny patients access to the necessary lifesaving medical
technology that is warranted. Again, Postman is silent on the real threat
to life—the denial of access to technology—in his militant determinist
case against overusing technology.

Postman (1992) attributes the intensive use of medical technology in
the United States to an American medical history of using the most "aggressive therapies available" (97). He illustrates his thesis with the
medical practice of Dr. Benjamin Rush, the famed eighteenth- and
early nineteenth-century American physician and his aggressive practice of bloodletting (96–97, 102–3). With a few minor variations,
Rush's treatment followed the traditional practices and "holistic" philosophy that dominated medical therapy until recent times and to which
many would have us return. The idea was that illness was a result of the
body's "humors" (blood, phlegm, black bile, and choler) getting out of
balance. Treatment was almost by necessity aggressive, as the physician's task was to restore balance by tactics such as bloodletting.

Tenner (1996) says, "Traditional medicine was like today's alternative therapies in viewing the patient as a whole person but in practice
was not necessarily gentle or humane" (39–40). Treatment was often
more lethal than the disease and had to be painful to be effective, or so
the doctors and patients believed. The discomfort and danger from
"bleeding, purging and mercury compounds were no mere inconven-
iences but part of the healing process, proof that the cure was working"
(Tenner 1996, 39–40). Calomel (mercurous chloride) and "the vomi-
tive tart emetic (antimony potassium tartrate) were administered so
frequently and liberally as to earn the epithet 'age of heroic therapy'
for the century preceding 1850" (Whorton 1974, 47).

Technologies, such as the stethoscope, and scientific understanding
such as germ theory, allowed medical practitioners to focus both upon
the specific cause of illness and specific measures to restore health.
They helped to reduce the sledgehammer mentality that dominated
treatments of the day. They are what Tenner (1992) calls "instruments
of localization," and they allowed "localizing surgery" and "localizing
pharmacy." Localizing surgery made it less invasive, and localizing
pharmaceuticals made directed treatment of the cause of the ailment
possible (41–44).

Localized therapy is not contradictory to a "holistic" approach to
treatment. Quite the contrary, the practitioner has to understand the
whole organism or person and the complex interaction of the parts in

order to devise a "localized" intervention. Few would argue against a greater concern for the "whole person" as part of medical practice, but it is naive and dangerous to believe that this concern is a substitute rather than a supplement to modern medical therapy. It is even more naive to believe that incorporating a more "holistic" (in the sense of treating the whole person and not as a New Age fad) approach would be going back to what was once "tried and true" rather than a continued forward movement in science and technology. A balanced look at the part and at the whole, an integration of analysis and synthesis, are what intelligent practice of any kind involves. What is being called "holistic" medicine in opposition to modern "reductionist" scientific medicine today may be emotionally satisfying to many, but it has yet to demonstrate a greater ability to further human health than by the medical knowledge it seeks to supersede.

Overusing Technologies

In many respects, it is understandable that technologies are "overused." Once accepted, some of the technologies discussed in this book border on being miracles compared to what preceded them. Their extraordinary successes in such areas as diagnosing illness, curing infectious diseases, and protecting food crops, are detailed later in discussions of food production, public health, and life expectancy. For example, antibiotics cured infectious diseases previously deemed incurable, so we thought we should administer them at the least symptom of infection. Chemical pesticides were cheap and effective, so we thought we should spray by the calendar, or whenever a few bugs appeared—and we didn't spare the chemicals. Of course, we now know that such practices lead to microorganisms or pests becoming resistant to the antibiotic or to the pesticide, but in the context of the early, extraordinary successes, they seemed to be very reasonable. And, in the case of antibiotics, it has been argued that the main reason we have problems with them today is that there is a long period from the late 1960s to very recently when no new class of antibiotics was developed because of the mistaken belief that we had all that we needed—we had closed "the book on infectious diseases" (Amyes 2000; May 2000).

The simple truth is that in the early stages of the use of a technology, we neither know all the potential beneficial uses, nor can we know the possible adverse outcomes. We learn by using good theories, by thinking through problems to possible outcomes, and by testing and experimenting in advance. To get a complete picture for any technology, particularly in the dynamic sense of technology in action, we must learn by doing. The ongoing process of the effective use of a stethoscope (or antibiotic or pesticide) is as much a part of the technology as is the physical instrument itself. To avoid misunderstandings—and misguided public policies—we have to look at the total process. Criticizing the overuse of a technology is a vital constructive part of understanding the technology and its potential. An approach such as that of the technological imperative leads to questions of technology being good or bad, gives rise to the wrong questions, and leads us nowhere. The answers to such questions can't tell us what actions are called for and what needs to be done, so the inquiry is academic in the pejorative sense of the term.

Intelligent criticism of the overuse of a technology asks the questions: How can we improve it and/or how can we continue to realize certain benefits while avoiding some of the previously unforeseen problems? These questions are part of the process and, I believe, the discourse involved in seeking answers to them facilitates technological change and the betterment of the human endeavor.

The difference between problem-solving criticism of technology and that of its opponents—technophobes—is not trivial. One simply cannot imagine any public policy or social issue that could be framed in terms of the technophobe's analysis.

Notes

1. Cited in Dyson (1999), 34.
2. The phrase "man makes himself" is from the title of a book by V. Gordon Childe (1951).

CHAPTER 2

Technology, Industrialization, and Creativity

S ome think that engineering and technology can be understood entirely as the application of scientific principles to practical problems in order to yield unique, deterministic solutions. Further, they believe that advances occur first in science and then are applied in engineering and technology. But there are many who object to these notions. First, it is clear that advances in technology have produced new knowledge in science as often as the reverse has been the case. There is a body of information that can be referred to as technological knowledge (Layton 1974, 31–34; Rosenberg 1982, 141–59; Sebestik 1983, 25–43; Ferguson 1992, 155–56). Engineers and historians of technology have demonstrated that many inventions had no preceding scientific or engineering principles that could have led to their creation. The idea to remember is that the historical relationship between art and technology that I discuss in this book continues up to the present. Those who criticize engineering for being "rote" or "mechanical" know nothing about engineering. I will explain why.

In a delightful article "The Mind's Eye: Nonverbal Thought in Technology," Eugene Ferguson (1977) argues that "all of our technology has a significant intellectual component that is both nonscientific and nonliterary" (827). Bridges are an example because they "reflect more strongly the conceptualization of their designers and the times of their construction than they do the physical requirements of their respective sites" (827). As in mathematics, there is "elegance" to technological solutions. They are so clear, simple, and obvious, that we all recognize the rightness and the solution to a problem (Ferguson 1978, 451; Hindle 1983, 128).

Among the illustrations for Ferguson's thesis is the working or rocking beam on the Newcomen steam engine, in which "the chain attached to the engine piston . . . and the chain attached to the water pumps in the mine . . . are constrained by the . . . curved ends of the working beam, to move in a vertical line, up and down, as the great beam rocks back and forth, like a see-saw" (451).[1] Anyone who has seen an illustration of the Newcomen engine, such as the one in Ferguson's article, immediately recognizes the curved ends as an elegant solution to a technical problem. The solution was "technically sweet." But as Ferguson muses, who among us would have thought of it beforehand, or what body of scientific principles would have led to this solution? In short, it reflects a creative vision, an intuition or an inspiration—as does the work of any other artist—even though its full implementation requires a knowledge of scientific and technical principles. As Ferguson (1977) concludes, "much of the creative thought of the designers of our technological world is nonverbal, not easily reducible to words; its language is an object or a picture or a visual image in the mind" (835).

There is a creative dimension to the machine. No writer has described it better than novelist Joyce Carol Oates (1973): "Machines are not evil, or even 'unnatural,' but simply extensions of the human brain. The designs for our machines are no less the product of our creative imaginations than are works of art, though it might be difficult for most people—especially artists—to acknowledge this" (38).

As Ferguson (1978) notes, "nobody denies the contributions of science to technology, but crucial non-scientific decisions are always involved in the designing or shaping of any technological object" (451). Some of the older technologies, such as that of the Gothic cathedrals, are sometimes romanticized and treated as achievements of the spirit. Billington and Mark (1984) don't agree: "Contrary to romantic wish, the great cathedrals took their glorious forms *because* their builders were forced to learn statics (empirically, of course), and their passion was channeled by that discipline into the creation of unprecedented forms" (39). Clearly, the inspiration was essential, as shown by Henry Adams (1963) in his classic masterpiece, *Mont-Saint-Michel and Chartres*. "The Archangel loved heights," he said in his opening sentence (1).

There was a powerful force motivating people to accomplish new feats of construction: the veneration of the Virgin. Adams (1918) says, "All the steam power in the world could not, like the Virgin, build Chartres" (388). It is equally true that all the inspiration in the world could not do it either, without the requisite technology. Or as the medieval builders would say, *ars sine scientia nihil est!*—art without science is nothing. Vision and technique are the essential qualities of all great engineering and architecture, past and present. Mark and Billington (1989) say, "The best engineer-designers did not devise building forms solely from applied science or out of any belief that efficiency and economy alone would lead to appropriate forms" (329).

Good engineering design combines "passion with discipline" with "aesthetic and the cultural meaning of their designs," a constant at the core of their discipline. Mark and Billington argue against the "easy" interpretation, which conceives of engineering "as merely an applied science" and believes that "only an architect can give modern structures aesthetic or cultural meaning." They add, "Outstanding modern engineers have succeeded in expressing the highest aspirations of our society by designing buildings with that indissoluble unity of structure and form in the same general way as did the Gothic master builders" (329).

The Industrial Revolution and the Arts

No previous transformation was as beneficial to human enterprise and creativity as the Industrial Revolution. Yet it was damned for being dehumanizing, and its technology was considered antithetical to artistic endeavors. The dualism between thought and practical action was revived with a vengeance.

Ambivalence and Democratization

It is more than appropriate that among the antitechnology artists and artisans, there was a revival of Greek forms in neoclassicism. Writer William Blake, who once referred to "dark Satanic Mills," was "dependent upon prosperous patrons for his livelihood" (Boime 1985, 111). The "fiery chariots," the furnaces, and other technologies were important images in Blake's poetry and drawings, reflecting more of an ambivalence to industrialization than is recognized by those who quote him. The neoclassicism in fine arts that followed Blake and the revival of Greek ideals were facilitated, according to Howard (1985), by "one of the first and most refined products of modern manufacture . . . the steel pen, which everyday recorded the images, means, and ideas of the new era" (790–92). Steel pens were better and cheaper than other writing instruments, and fell dramatically in price during Blake's lifetime. "The first commercially-produced steel pens cost $2.00–3.00; ten years later, they sold for about $.12; in 1823, for $.01–$.02, and after mid century, for less than a farthing" (794).

The steel pen was part of a larger process in the Industrial Revolution that facilitated "a new accessibility of artists' materials to the amateur market . . . formerly handmade by skilled craftsmen." Due to new technologies, they became standardized, manufactured, marketable products. Howard adds,

> Machine-made papers, grease and graphite pencils, steel pens, bottled inks and etching grounds, sticks of pastel chalk and watercolor, and premixed paints in tubes were easier, quicker, and more reliable to use than their forerunners and lent themselves to modern interests in clear, simplified, and brilliant effects (793).

Such new technologies advanced both the theory and practice of art and its public response, which included an increasing interest in color and compositional theory. An artist could spend more time making art rather than preparing materials. The new technologies: "encouraged more direct—even open-air—methods of painting and printmaking, which dispensed with tedious preparation of surfaces and materials, and, indeed, with the studio itself" (793).

This democratization of art by greater accessibility was occurring in other aesthetic activities. It was freeing artists from the need for patronage, and increasing affluence was creating a market for their creations. New technologies continued to provide new means for aesthetic creativity of all kinds, from music to painting to writing (DeGregori 1985, 70–73; Hall 1998, 218; Gettens and Stout 1966; Ong 1982, 94–95). The democratization of art brought on by industrialization was itself a condemnation to some who viewed mass products as inherently inferior and vulgar. As David Irwin (1972, 1976, 1988) comments, this view was "particularly misguided for several reasons, including that of historical inaccuracy." The Industrial Revolution "has also been a disseminator of good taste from the mid-eighteenth century onwards, a disseminator to all levels of society, and above all . . . one of the most important channels for the diffusion of the Neoclassical style" (Irwin 1976, 1988).[2]

Irwin adds, "in this industrial dissemination of good taste, painters, sculptors, and architects played a formative and influential role." An important criterion for assessing "taste," and technology's role in changing it, is not only what is, but also what was, replaced. In speaking of the "democratic ideal of American technology" and the many manufactured items produced for consumption, Ferguson (1979) adds perspective: "Rail, if you will, at the decline of taste; but look first at the real alternatives of bare walls, dirt floors, and minds untouched by the imaginative works of writers, poets, painters, and sculptors" (16).

Klingender (1968), a critic of the "debasement" of taste by mass production and marketing, nevertheless recognized that "the experience of the later 18th century . . . proves that, given certain conditions, industrial technique is not incompatible with the highest levels of design." Arthur Elton, the editor of Klingender's posthumous work, felt constrained to add,

were he revising his text today, he would have modified his views about the worth of Victorian taste and culture. . . . for there is now abundant evidence that, in the nineteenth century, mass production and the new industrial processes stimulated popular art forms no less vigorous and attractive than the "peasant" art preceding mechanization (47).

Elton uses the famed Crystal Palace as an example of the use of industrial technique that would have led Klingender to reevaluate his views of art and industrialization. He says,

With its soaring nave and transepts of iron and glass, the Crystal Palace of 1851, one of the great architectural monuments of Britain, was at once the product of mass production and of standardization, and a reflection of popular taste (47).

Above all, it was the democratizing force of the Industrial Revolution that allowed the "things of common use" to have a previously unattainable aesthetic value.

From the cast-iron detail of a station roof to locomotives, resplendent in brass and bright paint, from lamp standards to garden seats, from furniture to fabrics, from steel engravings to chromolithographs [they] often had a grace, delicacy and functional beauty Wedgwood himself would have admired (47).

Technology, Imperfections, and the Arts

With the advent of the Industrial Revolution, one of the qualities that made a craft item superior was its demonstrable inferiority. Through time, increasing precision was one characteristic artisans sought in refining and improving their craft. Nineteenth- and twentieth-century technology had carried this refinement beyond the point that hands or eyes could detect, and it had done it with mass production. Even today some people will point with pride to the imperfections that indicate handcrafting. Thorstein Veblen (1934) refers to the claims of excellence for a product that are based "on the degree of its approximation to the crudities" of earlier inferior technologies (121). Veblen was referring specifically to the very expensive books produced by William Morris and the Kelmscott Press. Veblen scathingly referred to the "painstaking crudeness and elaborate ineptitude" of Morris's process (122).

Veblen believed that limiting an edition—printing a small number of books—is the ultimate reversion to a criterion of earlier technology to enhance a product's pecuniary value (122). It is an artificially contrived scarcity, whether it be in the hand-produced item of William Morris or the mass-produced results of technologies. The crudeness must be contrived to be different because modern industry has taught us how to turn out great quantities of high-quality items (Kouwenhoven 1967, 35). It is a delicious irony that Veblen would truly savor were he alive today, that with computer assistance we can replicate precisely all the crudities of earlier works, such as the Book of Kells, except for the feel of parchment.

A slight modification of Veblen is in order. The artisan's continued refinement of his or her craft is still prized even as it approaches but does not quite attain the perfection achievable only by the machine. No "crudity" or imperfection is more esteemed than that in an object whose imperfection is so slight that it passes unnoticed to undiscerning eye and must be revealed to the observer by its proud owner.

The crafts of the Indians of the American Southwest in the late-nineteenth and early-twentieth century provide another example of less being more (Dilworth 1996). Mass-produced objects had a sameness to them, and because they were mass-produced they were by definition "perceived as being common and it is this commonness that the leisure class objected to." To Veblen, the leisure class "preference for the singular marks of imperfection, or marks of the hand, became 'honorific.'" For collectors of Indian crafts, "singularity and the mark of the maker's hand were highly valued" (Dilworth 1996, 154–55; Veblen 1934). In Chapter 1, I noted that the belief in the relationship of the hand and the mind of the Indian craftsmen marked them as being "survivals" from an earlier (presumably inferior) stage of human development, the "manual stage of cultural evolution" (Dilworth 1996, 151 and 154). The ironic twist is that the characteristics that could have made them inferior in the eyes of the buyer are the same ones that made their creations more valuable. It became a selling point. Typical of dealer ads for Indian arts and crafts was the following, which was on every page of a 1920 catalog: "Remember! Indians never make any two things alike" (155).

The Southwest thus became a source of singularity and authenticity. Tourists came to the Southwest "to experience 'firsthand' this authenticity" (155). Dilworth describes a celebration of technological simplicity in which the hand-brain relationship is ever present: "In making pottery the Indian uses neither measure, model nor potter's wheel. All is done from memory and with the hands. A few tools, hardly more than sticks and brushes made from yucca leaves, are the instruments" (Huckel 1913). And, "the Indian artist works without pattern, model—other than nature—and without rule or compass. The conception of the brain is brought directly to the place it is to occupy" (AOAI 1905).

Contemporary advertisers still use technological backwardness as a selling point for a product. Radios were touted as handcrafted long after printed circuits were proven to be more reliable, and one brewer still touted fire brewing when electric kettles were far better to control the heat. Some are even nostalgic for earlier technologies because they were less reliable: "The new cars are good, very good. They are almost perfect compared with what some of us grew up on. . . . But their predictability unreasonably makes us long for a past where every car and driver had a story" (Reeves 1997). And also, among those who were once competent to repair cars and other technologies, part of the nostalgia for less-reliable vehicles can be attributed to loss of the status—from being able to start someone else's car when they were helpless to do so.

In music, some enthusiasts like earlier forms of reproduction including the "warm tones" of vinyl (records). Some claim that digitalized sound reproduction causes headaches. With digital reproduction of prior recordings, music reviewers frequently praise new releases (or re-releases) for retaining imperfections that could easily have been removed. One reviewer states with reference to jazz recordings that they are not "just digital versions of vinyl LPs, these CDs were made from original masters and 78s; and despite noise-reduction processing, *they faithfully reproduce the background hiss of the earlier recordings*" (Birnbaum 1995). Denby (1996, 64, 66) has a "problem with perfection" and a preference for old recordings is also nostalgic about the lost hiss, while Holland (1999) pines for the "scratches and fluctuating turntable speeds to tell us" about a past that is being "denied, wiped away." Denby says, "Old recordings, with their slightly, honking bleat, their flattened high-frequency tones and muddied middle, can also be satisfying"(64, 66).

The characteristics being extolled in the old recordings are precisely the ones we sought to eliminate before the advent of the new technology. Prestige accrued to the committed audiophile with a very high-priced turntable (and not a proletarian changer) and other expensive, technologically superior equipment that produced a cleaner, more accurate sound, a task that was matched by studio engineers who sought to produce a recording for the discriminating listener. In other words, the audiophile devotees of "vinyl" want the latest, most advanced and best in the technology to which they are accustomed, but they reject a technology involving more than an incremental adjustment. Now that digital technology achieves the desired results with ease, the new elitist pines for and sings the praise of a sound once scorned. Technophiles and technophobes exist in many forms; some of us combine a bit of both.

Technology, Preservation, and Aesthetic Authenticity

Signs of aging and decay in buildings, furniture, and works of art have lent them prestige and are often maintained as marks of authenticity and antiquity, says Lowenthal (1985, 149–55). At times, the aura of aging has been deliberately contrived, sometimes for profit sometimes simply for art. In modern times, maintaining a condition of arrested decay for purposes of authenticating history and tourism in such enterprises as abandoned towns ("ghost towns") turns out to be a difficult technological achievement (Lowenthal 1985, 1979; Nelson 1977; Hart 1977). Lowenthal (1979) notes that preserved "structures are seldom visibly distinctive simply because they housed the famous; they are made distinctive only by means of markers."[3]

Decay gives another quantitative and often arbitrary dimension to authenticity. People who buy antique furniture wish to have it in good condition, which almost inevitably involves restoration. How much restoration is acceptable for Ming Dynasty furniture, for example? "Due to its age, wear and tear, most Ming furniture is damaged, and some degree of restoration is expected. Most experts agree that 10 percent is acceptable, but at 40 percent or more, the piece is deemed a fake" (Book 1997, 54). Appraisers on the PBS series "Antique Roadshow" repeatedly advise—to leave the defects. Removing the "patina" of cumulated dirt and grime substantially devalues the work even though it may improve its aesthetics.

Technology, Architecture, and Authenticity

John Ruskin, as much as William Morris, put forward canons of waste and ornamentation that Veblen so gleefully satirized. Ruskin (1880) defines art, particularly in regard to architecture, as that which "impresses on its form certain characters venerable or beautiful, but otherwise unnecessary" (16, 42–43, 44, 57-58). To Ruskin, building is a craft endeavor; architecture is adornment. Ruskin limits architecture to nonmetallic work, though he does concede that some day aesthetic principles suitable to architecture using metals will be devised. Further, all stones must be cut by hand. Ironically, the refinement in the use of metals and other materials allowed the ultimate in what Ruskin called adornment, for the structure of most modern buildings allows the load to be carried on a frame, the walls being structurally unnecessary curtains and therefore adornments, in Ruskin's use of the term.

In Ruskin's time and even before, there were already structures, such as the Crystal Palace, that achieved magnificence with steel and glass. Walls do keep out the elements and form separate rooms, but to achieve these objectives, the architect has many more aesthetic options. Further, in some construction, such as a suspension bridge or the Eiffel Tower, the designer can omit the adornment entirely, creating beauty and grandeur with pure functional structure. Instead of iron foundries and railroads, Ruskin would have had us build cathedrals (200–201). Ironically, it is the product of the foundries (and their successors) that gave architects in Ruskin's time and ever since the aesthetic possibilities never even imagined by previous builders.

The recognition of the enormous possibilities for beauty in unadorned structure is part of what Kouwenhoven (1967, 170) called the "energy of the vernacular" or the vital force that has emerged from popular culture and technology in industrial democratic societies such as the United States.[4] Structures made of modern materials achieve efficiency in both their creation and use. A reference at the famed Crystal Palace exhibit to American manufacture in general applies also to the best in modern architecture. "The expenditure of months or years of labor upon a single article, not to increase its intrinsic value, but solely to augment its cost or estimation as an object of virtue is not common in the United States" (17).

New materials (opposed by Ruskin) allow forms of construction, such as bridges, that are simultaneously achievements in engineering and art. It has often been said that it is almost impossible to conceive of an ugly suspension bridge. The industrial technology that created vast quantities of iron and, later, steel led to the creation of bridges that were beautiful, safe, efficient, and resource saving. Billington (1983, 30) writes, "early cast iron was about five times as strong as wood and hence required one-fifth the amount of material to carry the same load. The drastic reduction in quantity of material allowed the design to let more water flow past the bridge during a flood."[5] Prestressed concrete has given bridge builders and architects additional possibilities for creating structures that are both artistic and functional achievements (Copp and Zanella 1993, 197–241).

One of the most famous and appreciated unadorned structures is the Eiffel Tower in Paris. An iron tower three hundred meters tall, it was built by Gustave Eiffel to be the centerpiece of the 1889 Exposition Universelle and was the world's tallest structure at the time, says Gipe (1993). However, it was not appreciated by the "Paris literati, including famed writers Alexandre Dumas, Jr. and Guy de Maupassant" who "vociferously opposed the tower primarily on aesthetic grounds" (244).[6] A statement against it began: "We writers, painters, sculptors, and architects, lovers of the hitherto unspoiled beauty of Paris, protest with all our might and all our indignation" claiming to speak "in the name of unrecognized taste and in the name of menaced French art and history" (244). This bombastic proclamation stated their case "against erecting in the very heart of our capital the useless and monstrous Eiffel Tower, which the people's malice, frequently imbued with common sense, has already named the Tower of Babel" (244).[7]

A little over a century later, things had changed. When it appeared that the Eiffel Tower, or at least the mortgage, might be sold, the mayor of Paris sought help from the French government so that "the city's interests will be safeguarded" (Foster 1998). The French government assured the Parisian officials that the Eiffel Tower would never be sold.

The Myth of the Machine

Much of the rhetoric about machines controlling us is expressed by those who have not bothered to acquire the skills to control the machine. As Kouwenhoven (1967) argued in an illustration appropriate to his time, "the person who knows how to clean his own fuel pump filter and to adjust his ignition timing will be hard to convince that 'the machine' is his master." Today, many who, in their youth, reveled in tinkering with their cars—repairing, rebuilding, or customizing them—now bemoan the domination of modern life by digital technology in its many forms. Those young women and men, who have grown up with computers and an array of digital technologies that they have mastered, do not always share these concerns. I should not have to point out that the machine can be neither our master nor our servant unless we have plenty of those with intimate knowledge of differing machines that allows one to be comfortable with them, and of course, to keep them running (178).

Art Appreciation

There is no evidence that modern technology has diminished either the creation of art or its public appreciation in highly technological societies. In the United States, data regarding museum attendance, independent theater, and dance productions point to a growing enthusiasm for the arts. If critics of modern science and technology wish to argue a decline in the quality of contemporary arts, then they have a more subtle and difficult case to make.

Rasula (1990) says, "The postwar debate on mass culture reflected a troubling realization that unchecked circulation of high and low has the effect of transforming all culture into mass culture, regardless of who claims custody of it" (51). Some elitists argue that "mass culture" is a contradiction in terms. They attack the modern experience for the very success it has achieved in gaining access to art forms that have been virtually the esoteric domain of a select few.

Fundamentally, the critique of industrial and now postindustrial society is not about its failures but about its successes. In cultural matters, the term *popular* is interpreted aesthetically rather than literally. That is, "the adjective 'popular' has been utilized to describe not only those creations of expressive culture that actually had a large audience . . . but also and often primarily, those that had questionable artistic merit" (Levine, 1988, 31). Levine's critique of cultural elitism,[8] or what some of us would call snobbery, is amply illustrated in a later work that complains we are losing the distinction between aesthetics and vulgarity. "The gatekeepers can't find the gates," says Twitchell (1992).

The gatekeepers are necessarily the disinterested defenders of culture to which they lay claim. In an analysis equally applicable to developed countries, Orlando Patterson (1994a, 3–4; 1994b, 104) speaks of the "propagandistic reaction of traditional cultural gatekeepers in developing nations societies whose monopoly and influence have been threatened by the Western cultural impact." To Patterson, this cultural impact is generating new cultural forms and thereby stimulating the "emergence of new and competing cultural agents and managers. To monopolize the cultural resources of a country is to exercise power, not to mention to control economic resources." Most troubling to these elitists is that mass global culture leads to popular choice. Elitists across the political spectrum do not trust people to choose correctly.

Power of Music

Cultural diffusion has been a two-way street both within larger countries such as the United States and between the United States and other developed countries and the developing nations. The recording of music in the United States and the technology involved demonstrate the many ways cultural styles can interact and influence one another. Initially, recording was relatively expensive and confined to studio production. A few companies in New York basically controlled the industry, with "Tin Pan Alley" largely supplying the content (Hall 1998, 587–91).

In many respects, the sheet music industry, which arose with the mass production of pianos for middle-class households, took hold of the new technology of recording. By providing an outlet, these new recording technologies encouraged talented individuals to create an original and distinctive body of popular music, Tin Pan Alley. Many of the giants of this musical creation came from immigrant Jews who brought elements of their heritage to the music and made it American. In many ways their music was assimilated into mainstream American culture before they were.

By the 1920s, it appeared that this centralized production of music might someday lead to the demise of a significant number of regional, ethnic, and racial musical styles. Contrarily, by the late 1920s, new portable electronic recording technology allowed the major companies to go into the cities and towns of America and record the rich diversity of music that a multicultural, multiethnic, multiracial country spawned (Hall 1998, 571). The best way to sell records to people was to record their music. Advertisements were run in newspapers giving the date and location that the recording company would be auditioning local talent. One of the most famous of these sessions was in Bristol, Tennessee, in the summer of 1927, out of which came the founders of twentieth-century country music, Jimmy Rodgers and the Carter Family.

Further refinements in portable recording capability allowed folklorists and aficionados of diverse musical styles to literally go into cotton fields, churches, homes, prisons, honky-tonk bars, or anywhere music was performed to record the music heard there. Recording preserved an array of musical styles that would have been lost as various groups became more assimilated into the mainstream of American life. This preservation was helpful to music historians and was inspiration for the creation of exciting new musical forms. A classical example is the influence of the great 1930s blues artist Robert Johnson on the leading British rock musicians of the 1960s. The foundation was being laid for musical influences to move in more than one direction.

Tin Pan Alley's dominance in American music remained strong, fortified by other new technologies: radio and motion pictures. Jazz had broken out of its ghetto, though much of its performance remained segregated. Blues recordings were included in a genre called "race records" and had limited distribution. Performance was largely on small, local independent radio stations that were identified in terms of race. Still, for the blues and other forms, recording and radio meant that audiences outside the ethnic or racial group were slowly building and gaining an appreciation of variety in music.

In the 1940s, new technologies—television, LPs (long-playing records), multitrack recording, and FM radio—turned the trickle of influence of diverse musical styles, particularly African-American music, upon the mainstream of American music into a veritable flood (Hall, 1998, 584–91). The more expensive, centralized network programming moved from radio to television, creating an opening for more broadcasting of recorded music. From blues came R&B (rhythm and blues), then rock 'n' roll, or, as Muddy Waters sang, "Blues had a baby and they called it Rock 'n' Roll." They won their audience despite attempts to suppress them. The small production companies that had been recording blues artists and other newcomers suddenly found new audiences. They were able to challenge the previous tight control of musical production and dissemination by a few recording companies, broadcast networks, and ASCAP (the American Society of Composers, Authors and Publishers) (Hall 1998, 584–85). Markets worked the way they should, allowing new entrants to gain market share when those with monopoly power lose a sense of what the consumer wants.

The great music created by Tin Pan Alley still lives but now must compete with other sources of musical creativity. Technology did not deprive us of any music or musical inspiration, but, on the contrary, has created an extraordinary range of possibilities for musical creation and audience appreciation. The vitality that Patterson (1994a,b) describes burst forth not only with the crosscurrents of influence within the United States but also with the technology that has allowed rock 'n' roll to become a global phenomenon, producing new currents and crosscurrents with ever new and exciting possibilities that are revitalizing popular musical culture in many areas of the world. If we look at technology and popular music over the course of roughly a century, we can better see technology's potential for fostering diversity and creativity. This does not mean that we should be indifferent to current concerns about concentration in the production and distribution of music or worries about homogenization of culture, but that we should put such concerns in historical perspective as we work to continue to exploit the opportunities that technology engenders for sustaining creativity and diversity.

Cultural Diffusion

The global diffusion of the mass culture (pop music, film, etc.) that was largely created in the United States has been facilitated by the spread of literacy and technology, particularly communication, in former colonial countries. This diffusion has not led to a global homogenization of cultures, as many have feared, but has "generated enormous cultural production, in some cases amounting to hypercreativity in the popular cultures of the world" (Patterson 1994a, 3; 1994b, 104). True diversity is not what a mythical traveler from outer space would have encountered flying across the nineteenth-century landscape but what those living at the time, with the travel and communication technology of the time, would have encountered in their daily lives and throughout the course of their lives. Many lived their lives with a sameness and suspicion of any "alien" intrusion that we would consider intolerable today. So, rather than homogenization, cultural diffusion has provided choice and genuine diversity. Those of us who trust in human creativity believe that greater choice will lead to more and better cultural achievement, a belief more than vindicated by studies like Patterson's, which argue for the "increased vitality of local cultures and ethnicities in recent times" (Patterson, 1994b, 103).

Not only is there great diversity between nations and peoples, there is diversity within groups as the modern world increasingly reveals a vast range of styles and possibilities to those with creative impulses. One composer has referred to the United States as possessing a "mongrel culture" and meant it as praise for the opportunities that it offers creative talent. Lachiusa (1999) said, "We shouldn't have any one cultural tradition imposed on us. 'Pure' is a word that frightens me. We're mongrels in a mongrel country."

The critical distinction is not to let another tradition "exploit" you in the sense of being a slavish adherent to it, but rather "have the wherewithal to exploit it." The composer Michael John Lachiusa, proudly admits to being a "a cultural raider of sorts." He says, "I borrow what I want from European operatic tradition and I borrow from the American musical tradition. And from Balinese, Chinese, Native American, African musical and dramatic traditions."

The concern about mass culture and the influence of Western and American "global culture" derives from what Patterson calls "elitist postmodernist pessimism." The assumption of "postmodern" pessimists is that other peoples are "powerless" victims that unwillingly or unknowingly "passively accept cultural messages" and degrading cultural forms, abandoning their own (Patterson 1994a, 4). Quite the contrary, people are not merely sponges that absorb cultural forms but are active borrowers who create "further forms and varieties for everyday symbolic work and creativity" (Patterson 1994a, 4).[9]

Culture for the Masses?

When technology makes what was once restricted to elites available to the broad mass of a population, or even when a basic necessity that was once scarce becomes more abundant, elites will find reason to question the value of the technology or its outcome. Any cultural activity enjoyed by large numbers of people is presumed inferior. Intellectuals bemoan the fact that "people don't read" and then sneer at various mechanisms (such as the Book-of-the-Month Club) for mass marketing books. For them, mass culture is an oxymoron. Radaway (1992) sees the debate over the Book-of-the-Month Club as "an intraclass dispute over who precisely would stake the claim to cultural authority even while denying that this was being done" (523).

More than snobbery is involved in this aesthetic elitism. There are also questions of power. Power is a key ingredient in any number of issues discussed in this book. It is the power of elites to control what is read or at least to define what is worth reading. It is the power of elites to control the environment and lives of other people in the name of animal rights or saving nature. It is the power not only to define what is authentic in the art of one's own culture but also to define what is authentic and worthy in other cultures. And it is the power to exclude what is popular in any culture from having more than ephemeral entertainment value. It was the power of colonial and now extends to neocolonial elites to control the lives of others.

Lawrence Levine (1988) argues in *High Brow/Low Brow* that throughout most of the nineteenth century, American appreciation of Shakespeare was widespread, and what we now call classical music was performed on the same program with other music. In a word, Shakespeare and classical music were popular. After the Civil War, a number of changes began to take place in the United States. Immigration brought waves of people from different cultures, speaking different languages. Other technologies such as lithography and photography were making works of art accessible and large numbers of people could even own them. The elites felt threatened and needed to differentiate themselves from the lower orders of society; they had to reclaim their "heritage." Thus, there was the "cultural demotion of lithography" and the "cultural exclusion of photography" (160). Moreover, Shakespeare was to become the possession of the educated portion of society. Shakespeare was deemed to be "inaccessible" to the unwashed masses; performances should not be "compromised" to satisfy a mass audience (30, 70).

The means to limit access to high culture was not always pecuniary. They made it boring to the masses in the name of authenticity. Shakespeare became theatrical "spinach" (30). Levine says the primary debate was less over who should enter the art museum, symphony hall, or opera house, as over what they should experience once they did enter, "what the essential purpose of these temples of culture was in the first place. The 'fences and labels' . . . were firmly in place by the early decades of the twentieth century" (167). American literature, English literature of more recent centuries, and much of what is revered today as "canon" was attacked when it was introduced into the curriculum a century ago (Levine 1996).

The arts "were in effect 'rescued' from the marketplace, and therefore from the mixed audience and the presence of other cultural genres; they were removed from the pressures of every day economics and social life, and placed, significantly, in concert halls, opera houses, and museums that often resembled temples, to be perused, enjoyed and protected by the initiated—those who had the inclination, the leisure, and the knowledge to appreciate them" (230).

Many who "rescued" the arts from the marketplace were those whose fortunes were derived from the marketplace and whose wealth and leisure gave them the opportunity to experience the arts and control the experience of others. "There was one price that had to be paid: cultural products *had* to be accepted on the terms proffered by those who controlled the cultural institutions" (230–31). Restricting access is the "sacralization of culture. Culture is a process, not a fixed condition" (33).

Elitism and Being "Socially Responsible"

"Socially responsible" has become a catch-all phrase to describe a variety of well-intentioned, "feel-good" activities that presumably define those engaged in them as being "socially responsible" and maybe even morally superior to the rest of us. Money market funds are socially responsible. Collecting is "socially responsible"; shopping for eco-gifts is environmentally correct, and a long-distance phone service funds rain forest protection. Granted that some of these actions may result in the larger social good implied by the term. And many of these presumably "socially responsible" activities are simply harmless fetishes or affectations of the more affluent among us that produce neither social gain nor loss. But there are practices that are proclaimed to be socially responsible that are engaged in by those who have had the privilege of being raised in advanced scientific and technological societies. However well intentioned, their actions in the name of a "higher" ethical standard too often result in denying access to the benefits of modern technology for those in need. Nowhere is that more evident than in research in medicine and, currently, in food production.

Despite research achievements in public health for poor countries and the critical need for a greatly expanded research effort, there was harassment of a scientist, which adversely affected her research on "vaccines for malaria, schistosomiasis, giardiases, and other parasitic diseases that affect millions of people in developing countries" (Chui 1988, 1231). Currently, "less than 10% of the world's health budget is spent on conditions that account for 90% of global disease," with most of this disease burden being in the poorer countries of the world (GFHR 2000; Johnson and Van Dusen 1994). Drug companies are not only doing little research on pharmaceutical needs of Third World countries, they are often no longer producing those few helpful drugs that were available for tropical diseases. It is particularly heinous, then, to try to impede what little research is being carried out on the diseases that afflict the poorest and most vulnerable of the earth's inhabitants.

On New Year's Eve 2000 (December 31, 1999), vandals claiming to be the Earth Liberation Front torched Michigan State University's Agriculture Hall in order to destroy the offices of a researcher, Dr. Catherine Ives, director of Michigan State University's Agricultural Biotechnology Support Project, who was heading a project to increase food production in Africa and make it more nutritious (Kozlowski 2000; Andrews 2000). Africa is where the need is clearly greatest and progress has been far and away the slowest. For example, from 1970 to 1995, the percentage of malnourished children in the world declined by fifteen percentage points from 46.5 percent to 31 percent, while the absolute number declined from 204 million to 167 million. In Sub-

Saharan Africa the decline of four percentage points was overwhelmed by population growth, which has resulted in a 70 percent increase in the absolute number of malnourished children in Africa while the rest of the world continues to experience declines in the absolute numbers of malnourished children in addition to the percentage decline.

Those who act to disrupt vital research do not suffer the consequences if the research fails and one or more diseases remain virulent, nor will they go hungry if food production does not keep up with population growth. The "Report to Congress on the Extent and Effects of Domestic and International Terrorism on Animal Enterprises" by the U.S. Department of Justice found that from 1977 to June 30, 1993, there were "at least 313 acts" of terrorism against research involving the use of animals (Kaufman 1993, 1). Such attacks, the report says, peaked in frequency several years ago; however, "the violent nature of attacks had not diminished, with the majority of incidents being perpetuated at university research centers" (Kaufman1993, 1). The upsurge in animal rights terrorism in the United States has been matched recently by vandalism against genetically modified crops.

To those of us living in technologically developed countries, many of the health and longevity benefits we have are either the result of early childhood immunization (about which we generally had no personal choice), a more disease-free environment, clean water, better nutrition, or many other public health and medical benefits of a modern economy. Based on a 1985 study by the National Research Council (United States), Gross and Levitt (1994) found that 86 percent of the 139 Nobel Prize winners in physiology or medicine from 1901 to 1984 used animals in their research while the other 14 percent used plants or microorganisms (200–201). Their research and that of others using similar means have been an essential element in the extraordinary twentieth-century advances in life expectancies and human well-being.

We are all passive beneficiaries of the efforts of others. Generally, we are unaware who our benefactors were or the almost unbelievable extent to which our lives were improved and extended in the twentieth century alone. In this regard, those who violently oppose scientific research and others who more peacefully oppose scientific and technological change can accept the benefits of this research and development without violating their principles only by remaining unaware of the source of their betterment. Nevertheless, the data are so huge that it is hard for the technophobes to deny the benefit, so they simply deny the reality. They still must justify their actions that would deny these benefits to others. While the animal rights groups oppose animal tests, other purist environmental groups make demands for testing of almost everything new for safety, which requires more animal testing, not less. Modern science has been able to devise some new techniques, often using computer simulation, which avoid or minimize the use of animals, but it is a clear and overwhelming consensus of the research community that animal testing remains an essential component of a wide range of research activities. The public, acting through government agencies, has placed a variety of constraints upon the use of animals in research so that whenever animals are used in research, their use is as humane as possible and necessary for a research endeavor with a potential outcome that has benefit for human well-being that is sufficient to warrant the use of animals. Scientists have largely accepted these constraints though some complain that they are too restrictive.

Even if one is willing voluntarily to forgo these benefits, which few are likely to do except in the most superficial way, there is still no justification in forcibly acting to deny them to others. For the more fervent believers in earth first or animal rights, there are many benefits that they can voluntarily forgo. They can, if they wish to be honest and consistent, carry the following card that has been distributed by those who favor animal research. On one side the card reads as follows: "I hereby identify myself as a supporter of animal rights and agree to live my life in accordance with all animal rights principles." On the other side of the card it says,

> So as not to violate my animal rights principles, I hereby request that in the event of an accident or illness, all medical treatments developed or tested on animals be withheld, including, but not limited to: blood transfusions, anesthesia, pain killers, antibiotics, insulin, vaccines, chemotherapy, CPR, coronary bypass surgery, reconstructive surgery, orthopedic surgery, etc.

The various vaccines used in modern immunization are overwhelmingly the result of research using animals. If immunization has not saved more lives this past century than any other intervention, it certainly has to rank among the top two or three life-saving and life-sustaining developments. The much-maligned chemical pesticides have also played a role in saving lives during the twentieth century even though to many, the mere evocation of the term evokes an adverse reaction as a threat to humans and to life on the planet.

Immunization has its many triumphs, as in smallpox eradication, but pesticides take credit for the battle being won against onchocerciasis, a disease caused by a parasitic nematode (*Onchocerca volvulus*) that results in excruciating pain and permanent blindness in humans (river blindness) in West Africa and other tropical areas. From the World Health Organization (WHO) (1999c) website on onchocerciasis:

> It gives rise to serious visual impairment, including blindness; to intensely itching rashes, wrinkling and depigmentation of the skin; to lymphadenitis, resulting in hanging groins and elephantiasis of the genitals; and to general debilitation.

However horrible the above description sounds to those of us raised in societies with few infectious diseases, the reality is even worse; one can imagine the horror, debilitation and pain caused by as many as one hundred million microfilaria working their way through one's body, feeding on nutrient vital to the person and long, 30–80 cm for females, adult worms lodging in connective and muscular tissues or in fibrous nodules in one's skin (Burnham 1998, 1347; WHO 1995, 3). The infecting bite of the black fly vector will produce an ulcerated lesion with an almost unbearable pruritus and scratching that last for weeks (Nwoke 1990, 38). Previously, 60 percent of the population in areas of West Africa were infected, and 3 to 5 percent became blind (Liese 1995, 11).

The WHO's Onchocerciasis Control Program used a combination of pesticides (sprayed by helicopters) and ivermectin for treatment, dramatically reducing the scourge of this disease (Pud'Hom et al. 1993). Satellites were used to transmit hydrological data (on rainfall, run-off, soil humidity, river and stream levels, soil moisture, humidity, etc.), which were useful for spraying decisions (Servat et al. 1990). "The larvicides applied in OCP in rotation to avoid resistance are temephos, phoxim, pyraclofos, B.t. H_14, permethrin, carbosulfan and etofenprox" (Christensen 1999; Molyneux 1995, 400; Calamari et al. 1998, 485). Of these, temephos, phoxim, and pyraclofos are organophosphates, permethrin and etofenprox are synthetic pyrethroids, and carbosulfan is a carbamate (Hougard et al. 1997, 425; Hougard et al. 1993, 436). Continuous, careful monitoring of the areas sprayed found "no long-term impact of pesticides on the environment" (Molyneux and Morel 1998, 332).

Prior to the program, about 120 million people in thirty-seven countries in Africa, Latin America, and Yemen were at risk for the disease, with roughly 18 million people infected with onchocerciasis and over a million people with impaired vision, a quarter of whom were blind (Garate et al. 1996, 490). Most of the cases were in West Africa (Remme 1995, 403). By 1995 in West Africa, over 30 million people in eleven countries "were protected from damaging ocular lesions," 10 million children who were born since the program began were "spared the risk of blindness," and 1.5 million people "originally infected are no longer so" as a result of the Onchocerciasis Control Program (Molyneux 1995, 400).

In February 1999, the WHO declared the near eradication of river blindness in West Africa, though the black fly that transmits the parasite has not been eliminated. In addition to lives saved and people protected, up to "25 million hectares of land that could feed 17 million people using traditional methods of cultivation were opened." In strict economic terms, the increased agricultural productivity represented a "rate of return on investment of around 20% which far exceeds that associated with health projects" (Molyneux, 1995, 399). It is estimated that the eradication program cost $437 million (1985 dollars), which works out to about $0.53 to $0.57 per person protected per year (Kim and Benton 1995, 3; Molyneux and Davies 1997). Efforts are underway to eliminate onchocerciasis in the Americas by 2007 and globally by 2010 (WHO 1996, 277; REB 1996; Hougard et al. 1998).

Prior to the West African eradication program, onchocerciasis was successfully eradicated in Kenya in the late 1940s and in the Mbari forest of Uganda in the 1960s. This was done largely by spraying with DDT (Fischer et al., 1997). From the mid-1940s to the early 1970s, DDT was the insecticide of choice to control and/or eradicate onchocerciasis wherever it was endemic (Davies 1994, 24-35). The program in Kenya cost between $1.82 to $3.35 per square kilometer, which is incredibly cheap, even in 1940s dollars.

Other horrible diseases unknown to us in temperate-zone developed countries are being dramatically reduced and are even on the verge of eradication through programs similar to the onchocerciasis project using a combination of pesticides and medication. Ivermectin is also vital to a campaign to eradicate lymphatic filariasis (WHO 1999d). "There are 120 million people with this tropical disease and a further 1.2 billion—one fifth of the world's population—at risk" (Dean 2000). Dean says, of those infected, "76 million have pre-clinical disease, with lymphatic and renal damage; 44 million have recurrent infections. Up to 15% develop lymphoedema; another 10% elephantiasis; and up to 50% of men develop the dreaded testicular hydrocele." Progress is also being made on the eradication of another scourge of West Africa, guinea worm, with a 97 percent reduction in the disease (dracunculiasis), and in India it has been eliminated (Peries and Cairncross 1997; WHO 2000c; Sharma 2000).

The Eradication of Smallpox

The adage "I can take care of my enemies; Lord, save me from my friends," applies to many who seek to save other people from the benefits of technology in the name of respect for another culture. Nandy and Visvanathan (1990) maintain that Western technology was not the reason for declining death rates and that people in developing areas would do better by using their traditional medical technology. First, they argue that death-rate declines in Western countries resulted from improved public health, without their realizing that public health improvements such as clean water were the products of science and technology (148).

Most expert demographers agree that public health measures were the primary determinants of death declines to the 1930s, even to 1950. But many include immunization as a public health technology. Nandy and Visvanathan add the argument that the major diseases were evolving toward being less virulent. This is a theory that once had wide acceptance, but it fails to explain why the same diseases were evolving to become less virulent in developed countries but not in poorer countries. Nor does it explain why these diseases suddenly became dramatically less virulent in developing countries following the introduction of programs such as immunization, oral rehydration therapy, or vitamin A intervention. Paul Ewald (1994) argues:

> Few ideas have been so ingrained in the literature of medicine and parasitology as the idea that parasites should evolve toward benign coexistence with their hosts. Few ideas in science have been so widely accepted with so little evidence (3).

Ewald recognizes that some diseases do become less virulent over time, but also that some become more virulent. In other words, evolution of disease vectors is complex, not strictly unidirectional, and therefore one cannot use the "idea that parasites should evolve toward benign coexistence" to explain declining death rates in developed and developing countries.

Immunization has been one of the most important interventions that revolutionized life in the twentieth century. This is reflected in the fact that those who had privileged access to medical care a century ago were able to gain little benefit from it. In the United States in the 1890s,

> professionals did not have child mortality levels that were very different from those of other groups, and literate classes enjoyed less of an advantage than they do today. Particularly telling is the mortality of offspring of doctors, which was only 6 percent below the national average (Preston and Haines 1991, 209).

Surely no one could oppose immunization in the twentieth century. Belief in a harmonious nature and a natural evolution of microorganism to being relatively benign can contribute to such a belief that immunization is unnatural, unnecessary, and harmful, but not fully justify it. Prior to immunization, the main prophylaxis against diseases such as smallpox was inoculation. In any human action, there are risks, even in those activities that are otherwise lifesaving. Whatever the risks, Bedford and Elliman (2000) state, "immunization against infectious disease has probably saved more lives than any other public health intervention, apart from the provision of clean water." The risks of side effects from smallpox immunization in the United States were 0.00011 percent from all causes (both direct death or from complications). This risk factor was twenty thousand times less risky than inoculation, itself considerably less risky than no protection at all (Copp and Zanella 1993, 259–60).

Even for the controversial pertussis vaccine, there were and remain clear benefits.

> Up to 250,000 people per year, mostly children, suffered from pertussis in the 1930s, and approximately 7,000 people died each year in the United States . . . when the DTP combined vaccine was . . . put into routine use in the United States, the case rate soon fell by a factor of 100 to approximately 2,000 with fewer than a dozen deaths each year (Copp and Zanella 1993, 276–77).

The increased death rate from pertussis that followed the decline in immunization in England and Wales is further evidence for the life-saving benefit of immunization. The *alleged* complications from the vaccine that led many not to have their children immunized against pertussis have been corrected in a vaccine that was recently tested and is now being used. Given the complexity of the epidemiological issues involved, it is also possible that the observed complications may not have been caused by the pertussis vaccine but by a complication that tends to strike at the same age at which the vaccine is normally received.

What is clear is that the warnings about the dangers of the pertussis vaccine did cause a loss in life for children of those who were alarmed and did not have their children immunized. The lesson is that warnings of danger are not risk free, and even when scientifically based, warnings need to be very carefully phrased, and the media should exercise additional care, lest they help perpetuate a false alarm. In early 1998, there was another vaccine scare in the United Kingdom; this time it was the vaccine for measles, mumps, and rubella. The scientists for the Medical Research Council investigating the issue found no verifiable links between immunization and autism and other maladies. They did note the verifiable dangers of not being vaccinated, as did Finnish scientists who gave "the controversial triple vaccination a clean bill of health" (Lancet 1999a, 2000c; AAP 2000; BBC 1998a).

Even if someone distrusted immunization, despite its enormous successes, one would think that there would be no objections to the total elimination in ten years and nine months (January 1967 to October 1977 with victory being declared in May 1980) of a disease—smallpox in this case—that was infecting 10 to 15 million people and killing 1.5 to 2 million people each year. At the time the eradication campaign began, endemic smallpox had already been eliminated from 125 of 156 countries (Fenner et al. 1988, 175). During the first eight decades of the twentieth century, from 200 to 300 million people died from smallpox, two to three times greater than the estimated 100 million deaths caused by armed conflict (Henderson 1996, 39; Oldstone 1998, 27; Koprowski and Oldstone 1996a, 19). It is estimated that prior to immunization in the eighteenth century, "nearly one-tenth of all mankind had been killed, crippled or disfigured by smallpox" (Koprowski and Oldstone 1996b, 23).[10]

In 1967 when the program began, the World Bank (1993) says

Somewhere between 1.5 million and 2 million people died of smallpox. Perhaps half a million more were blinded, and more than 10 million were seriously and permanently disfigured. In the early 1950s the toll had been three or four times greater (17).

In the 1950s, a number of countries had begun their own eradication campaigns, so when the global program began, smallpox had been virtually eradicated in 125 countries. By 1950, smallpox had been eradicated in North America and Europe as vaccination had begun in the early 1800s (Crosby 1993, 1012). Within five years (1972) of the start of the eradication campaign, smallpox no longer existed in the Western Hemisphere, and by the end of another year (1973), smallpox "was restricted to the Indian subcontinent and the horn of Africa, Ethiopia, and Somalia" (Crosby 1993, 1012; Oldstone 1998, 43–44).

The benefits of the eradication program have been monetary as well as humanitarian. The cost of smallpox vaccination, quarantine programs, and treatment totaled more than $300 million in 1968 alone. The eradication program, by contrast, cost $300 million over its twelve-year life. It has therefore saved hundreds of millions of dollars a year in direct, measurable costs (World Bank 1993, 17).

There was considerable economic benefit from the smallpox eradication campaign even for those countries free of smallpox, since it "enabled the discontinuation of prevention and treatment efforts, including routine vaccination." By 1985, the United States recouped its investment in worldwide eradication "every 26 days" (CDC 1999b, 393; MMWR 1999b).

Eradicating Smallpox: A Mistake?

In 1990, two scholars argued against the eradication program saying, "smallpox need not have been eradicated; it could have been contained" (F. A. Marglin 1990, 140; S. A. Marglin 1990, 20). One of the scholars, F. A. Marglin, deconstructs the science from which immunization is derived and finds it guilty of "logocentricism" and "phallogocentrism." In India (and elsewhere), smallpox was historically contained by a method called variolation and "*only* about 1 in 100 persons died even during epidemic phases of the disease."[11] Variolation involved taking the pus from a smallpox victim and directly inserting it into an opened wound of the person to be "protected" by it (Fenner 1996, 33–34; Fenner et al. 1988, 245–58).

Variolation was intertwined with the worship of Sitala, the Hindu deity for smallpox. Marglin argues that variolators, with the proper education, could have been used as vaccinators. "The aim of vaccinating 80 percent of the population could have been attained if the indigenous system of variolation had not been destroyed" (140). This "would have been sufficient to prevent devastating epidemics from occurring" (140).

With Marglin's plan, the disease would have remained and continued to kill people, but presumably *only* an acceptable few. Marglin is wrong to claim that the smallpox eradication program did not use variolators as vaccinators. Everywhere, most definitely including India, the program worked closely with governments at all levels, using whatever local resources were available (Henderson 1994, 232; Fenner et al. 1988, 711–91). In India, a smallpox eradication campaign was long in place prior to the commencement of the global effort and provided an essential foundation for India's eradication program. Those involved in the eradication campaign were very much aware of the worship of Sitala and sensitive to the issue of the saints and deities in many religions and cultures from whom protection from smallpox was once sought (Fenner et al. 1988, 219–23). India had already made significant inroads, reducing mortality by the twentieth century from a nineteenth-century vaccination program (Banthia and Dyson 1999).

There are more than a few inconvenient problems with F. A. Marglin. Containment means people continue to die from smallpox. There would be no big epidemics but plenty of death and blindness. After all, 20 percent of India's current population would mean 200 million people (and growing) at risk from smallpox. Who would choose those to be vaccinated and those who would not?

What if more than 80 percent of the population wished to be immunized? How do we deconstruct that reality? How do we show respect for the wishes of that culture? Epidemics would be "contained" only for populations that were largely immunized. The world's other five billion plus people would forever have to be immunized (with a cost in lives lost, directly and indirectly, because of health resources diverted from other uses) until a variety of smallpox emerged that was resistant to the vaccine. Since the eradication campaign did not use universal immunization but devised an effective strategy of "identification and containment," then a residual pool of smallpox existing in India would have immediately put many hundreds of millions of people around the world at risk. Smallpox immunity begins to decline after about twenty years, so the vast majority of the world's population today, including those previously immunized, would theoretically be vulnerable to smallpox infection.

All this is advocated in the name of respect for the beliefs of another culture. To Marglin, the democratically elected government of India has never experienced "decolonization" of its mind (F. A. Marglin, 1990, 140). Fortunately, the government of India has a profound understanding of its own beliefs, the rights of its own people, and the people of the rest of the world.

The Marglins and their deconstructionist kith and kin are guilty of the same sin they attribute to modern science: namely, they are convinced of the absolute superiority of their system of belief and analysis. India has long prided itself on having the third largest cadre of scientists and technologists (after the United States and what was then the Soviet Union). One wonders what distinguished Indian scientists and technologists in India and around the world would think of those who view their achievements as being the product of colonized minds. Most would fiercely deny the implicit thesis that modern science and technology are a unique, exclusive product of Western culture.

Our academic guardians of postmodernism are guilty of "orientalism" (the presumption of outsiders that they can inform a non-Western people what their true values are) and are more ethnocentric than those whom they criticize. They presume to be more authentic spokespersons for India's cultural traditions than are hundreds of thousands of its people and government. In another context, Tomlinson (1991) argues that the "temptation is strong for intellectuals who do feel cultural imperialism as a threat to 'speak for' the culture by attributing a form of 'false consciousness' to the masses who don't" (94).

Presuming to speak for the poor peoples of the world has become a disease of epidemic proportions among advocacy groups in affluent countries who are promoting a policy agenda that is in fact counter to that being sought by the leaders of developing nations countries. Few issues so frustrate and infuriate developing nations' leaders as the actions of organized environmental groups who use their influence to block loans and interfere with their development efforts and then profess a moral superiority for doing so.

Indira Gandhi commented, "How can we speak to those who live in villages and slums about keeping the oceans, the rivers and air clean when their own lives are contaminated at the source? The environment cannot be improved in conditions of poverty. Nor can poverty be eradicated without the use of science and technology" (Jukes 1974, 15).

Notes

1. See also Ferguson (1992), 14.

2. See also Irwin (1972), 288–97, and Boime (1987).

3. He adds, "instead of rescuing history from obscurity, such markers drown it in trivia." Thus, "to realize something stems from the past is actively to alter it" (Lowenthal 1979, 109, and 110, see also 116–25; and Sopher 1979).

4. See also Jackson (1984) as well as other works by Jackson.

5. On the improvements and aesthetic significance see Gordon (1988).

6. The ultimate insult was from the composer Charles Gounod who argued that "even commercial America wouldn't want the Eiffel Tower" (Salinger 1989, quoted in Gipe 1993, 244).

7. Vidal (1988) quoted in Gipe (1993).

8. The words *elite, elitist,* or *elitism* are here defined in context as those who believe that only a small minority (an elite, if you prefer) are capable of appreciating high culture and therefore anything popular is thereby inferior. With specific relevance to this book, antitechnology elitism is opposition to technologies that "mass produce" or otherwise make available to a larger audience what some consider should be restricted to an elite.

9. See also Willis (1990), 26.

10. Smallpox sufferers "developed spots like flea-bites, which grew into pustules containing a transparent fluid which turned into a thick pus. The eyelids would swell and become glued together. Sufferers had to be prevented from tearing their flesh to shreds" (Kedzierski 1992, 1).

11. F. A. Marglin (1990), 110.

CHAPTER 3

Technology, Technophobia, and Human Welfare

In the early fall of 1998, there was a story on the news services about a statistics teacher who had his students do an assignment of either flipping a coin one hundred times and recording the results or making up the entire assignment. He could always tell which was which by the fact that those who made up the assignment failed to have the runs of heads or tails that would be expected as the random outcome of flipping the coin. Simply stated, *random does not mean uniform*. This is a point that needs to be repeated over and over again.

Cancer Clusters

There is a growing concern in the United States over "clusters" of cancer and/or birth defects (Gawande 1999). As would be expected, the immediate suspicion is that there are hidden chemicals (meaning man-made chemicals) in the environment that are the cause of the cluster (Sachs 1998). In epidemiology, clusters of a disease often become the basis for investigation into a possible local cause. Obviously, a cluster that is several hundred or even several thousand times what would be expected in a random distribution affords more than sufficient reason to initiate a thorough epidemiological investigation. Over the years, some of these inquiries have discovered disease vectors and environmental causes both man-made and natural (Gawande 1999). Current concerns about clusters and "cancer mapping," however, are, at best, well within a range that would be expected by random variation. In fact, given the number of different types of cancers, it has been estimated (for the state of California) that any particular census tract would have a "better than even" chance of having "a statistically significant but perfectly random" elevated level of a particular type of cancer (Gawande 1999, 37).

A *Civil Action* is a movie based on the premise that clusters of a cancer must be caused by a chemical. Gladwell (1999, 23) and Jaeger and Weiss (1993) are critical of *A Civil Action* arguing that on its "scientific premise," the film "permits no ambiguity at all." Further, there is no evidence that the "chemical allegedly dumped, trichloroethylene (TCE), is a human carcinogen." It is a "probable human carcinogen" based on animal tests, but "no human-based data have tied it to cancer." The film failed to mention that there were high concentrations of other chemicals, chromium and arsenic, in the same watershed from much earlier contamination (Reisch 1999, 20).

Resources are being allocated to find these alleged chemical carcinogens, even though many scientists are skeptical and believe that the resources could be better spent in research or some other public health endeavor. Some scientists argue that, with the many types of cancers and birth defects that are possible, clusters in small communities are almost inevitable (Kolata 1999a). Add the fact that there are tens of thousands of different chemicals manufactured each year, and it would be possible to find a chemical or chemicals that are associated with a cluster without their necessarily being the causal agent of any human malady whatsoever.

Hysteria and Random Variation

The hysteria that arises from fears of chemical toxins or new viruses has been compared by Wessely (2000) to earlier outbreaks of witchcraft hysteria. "In a previous era, spirits and demons oppressed us. Although they have been replaced by our contemporary concern about invisible viruses, chemicals, and toxins, the mechanisms of contagious fear remain the same." For the people who experience this phenomenon, called "psychogenic illness," the physical symptoms are real and often long lasting, making it understandably difficult for them to admit their problems are psychological in origin (Jones et al., 2000; Wessely 2000).

It is difficult for public health professionals to respond in a manner that is intelligent and sensitive to the needs of those experiencing psychogenic illness, while trying to educate the public on the psychological origins of the problem so as to prevent or at least minimize the spread of this contagion. Those who have made careers out of spreading fears about chemicals bear no small responsibility for these outbreaks of hysteria, the pain that they cause, the public health resources they waste, and the legitimate environmental problems that may then not properly be evaluated and addressed.

Searching for causal agents where there might not be any is potentially a bottomless sump, absorbing scarce public health resources. One must consider the benefits of scientific research and public health programs and the improvements in our lives that they have brought, a point that I return to repeatedly throughout this book. If I had a cynical belief that all public expenditures of any kind were a waste, then this particular waste would not bother me. The real cost of wasted resources is the lives that suffer as a result.

An argument by public health officials about random variation is seldom convincing to citizens whose children are dying of cancer. Gawande (1999) says, "How many more of our children must die before something is done?" (37). We seldom think of how many more will die when public health resources are diverted from more productive uses.

The causes of cancer could be genetic, diet, lifestyle, or any number of possibilities other than environmental chemicals. Virtually ignored in the public discussion of clusters in the United States is the reputable epidemiological thesis of Leo Kinlen that clusters of one type of cancer, childhood leukemia, may be the result of a virus and varying immunities from population mixing (Kinlen 1988, 1989, 1992, 1995, 1997a,b, 1998; Kinlen and Stiller 1993). Recent epidemiological work appears to verify Kinlen's thesis (Dickinson and Parker 1999; Doll 1999). The public health discourse on presumed clusters needs to consider Kinlen's work and not let it be drowned out by the cacophony of chemophobia.

Causes of Clusters

Any statistician can verify that random distribution does not mean uniform or even distribution but can create local clusters without the necessity of there being a local cause, chemical or otherwise. The mere existence of a cluster tells us little without knowing the size of the cluster and its statistical probability or improbability and without some scientific, statistical, and epidemiological inquiry. If one is living in a cluster area and is either afflicted with the malady or has an afflicted loved one, it is reasonable to believe that there is a causal agent for the cluster. It is reasonable to believe that; unfortunately, it is wrong. We cannot say a priori that any cluster is not caused by an environmental chemical. The issue is how to deal intelligently with the question of clusters. It is virtually certain that some basic epidemiological investigation will be conducted whenever clusters emerge. The critical matter is how many resources to devote to this type of task and whether the same resources directed elsewhere could lead to a greater expansion in human health and well-being. It is also important for the media and others who influence public perceptions to educate the public about some of the basic issues for understanding the distribution of various afflictions and to do so before the emergence of clusters might bring in emotional factors that cloud good judgment.

Science, Public Policy, and the Scientific Method

Any study that purports to find that a manufactured chemical, particularly a pesticide, may cause cancer or other human disorders gets prompt, massive, media attention. When certain chemicals in the environment acting synergistically were found to be estrogenic, possibly leading to an increase in breast cancer, it was widely publicized, but when this report was withdrawn because neither the original researchers nor anyone else has been able to replicate the reported results, I could find no media coverage (Arnold et al. 1996). The letter withdrawing the report states:

> People in many walks of life have, on their own, put great weight on this report as the basis for much discussion, thought and public policy. Whatever merit this publication contained, and despite the enthusiasm it generated, it is clear that any conclusions drawn from this paper must be suspended until such time, if ever, the data can be substantiated (McLachlan 1997, 463).

Unfortunately, public policy was based upon this report with continuing impact up to this writing. Its findings found their way into the Food Quality Protection Act of 1996 (discussed in Chapter 5), which includes a directive to the EPA (Environmental Protection Agency) to study the impact of the estrogenic compounds, or endocrine disrupters as they came to be known. This study found no epidemiological evidence nor could it define any biological mechanism by which the purported disruption took place. Because the environmentalists were unhappy with the null results, the latest study calls for further study (NAS 1999; NRC 1999a).

Flawed Studies and Funding Phobias

The problem is that, once a causal connection between a phenomenon, such as endocrine disrupters, and a human disorder is posited, an academic industry is generated: grants and research institutes are created and newly minted experts become media pundits on the subject. This constitutes a powerful force that is resistant to any research findings contradictory to the previously established presumption of danger. Any study refuting an alleged environmental danger is deemed to be fatally flawed. The real fatality may not be the new study but some cherished assumptions. For example, when a recent study found no evidence of any link between the Three Mile Island nuclear power accident and cancer deaths, an antinuclear power activist immediately labeled the study to be flawed and a recitation of the industry line (Strawley 2000). What is interesting about the swift reaction was that the full study was not yet available, only an abstract posted on a journal website (Talbott et al. 2000).

Ironically, like the withdrawn study on the estrogenic compounds, the only studies on EMF (Electro-Magnetic Fields) that appear to be flawed are some of the early ones claiming a relationship between electromagnetic fields and cancer. The laboratory that conducted some of these studies has conceded that some of the data were fabricated, leading the U.S. government to request that partial research funds be returned (Vergano 1999; Liburdy 1999). The web version of Liburdy's letter to *Science* defending his work has a hyperlink to an EMF website where prominently featured is a plea for letters to Congress calling for continued funding for EMF research, along with a draft letter and other helpful hints (Bioelectromagnetics Society website www.bioelectromagnetics.org—for letters, add /congress.html.). The draft letter includes the following: "A 20-year Federal investment in scientific studies to determine possible health effects of environmental electromagnetic fields is in jeopardy."

This is saying, in effect, that we should continue to waste money on fruitless research because we have been doing so for twenty years. The EMF report itself found the evidence for the thesis is weak, the probability of health hazard is "currently small," and "the evidence does not seem to meet the standard for listing as a known or even 'anticipated' human carcinogen in the National Toxicology Program's Report on Carcinogens." Yet there was a call for continued funding, though at a reduced level (NIEHS 1999). Since total, absolute proof of no risk is essentially impossible, once a phobia is funded, it takes on a funding life of its own.

These phobias, then, are not costless. The research pursuing them takes valuable human and financial resources that could be used in more productive research. By 1992, the annual cost to society of the EMF phobia was estimated to be over $1 billion a year (Florig 1992). Despite the mounting evidence against the EMF-cancer connection, a 1999 survey commissioned by the Cancer Research Campaign in the United Kingdom found that over half of those surveyed accepted the EMF-cancer connection as fact and as part of a complex of other myths about cancer and its causes (NRC 1996b, 1997; Linet et al. 1997). Campion (1997) describes the effort to find a connection between electric power lines and cancer: "It is sad that hundreds of millions of dollars have gone into studies that never had much promise of finding a way to prevent the tragedy of cancer in children" (46).

In addition to the hundreds of million of dollars wasted on EMF studies, the White House Science Office estimates that the total cost for this fiasco is in excess of $25 billion (Park 2000a, 161; 2000b). This figure includes relocating power lines and the loss of property values but obviously does not include the human anguish of those who were led to believe that they had put their loved ones in harm's way by living near power lines or allowing their children to go school in close proximity to them. Those who were once thought to be the innocent victims of living near power lines became the victims of those who sought to save them.

Saying It Makes It So

Once a chemical or other product is labeled a carcinogen or dangerous to humans in other ways, it is almost impossible for scientists to deem it to be safe, even if they are unable to find evidence of its harm (Kolata 1999d; NAS 1999; NRC 1999a). Nor can they avoid calling for further studies. The first step in the process of condemning a product as a carcinogen or dangerous in other ways is to give it a frightening name that assumes casual relation between a chemical or practice and a horrible outcome that is yet to be proved. Unfortunately, the name remains, condemning the product long after scientific investigation has found no such relationship. Once again, not having evidence is no problem, as Theo Colborn, a proponent of the endocrine disrupter thesis, simply stated: "this is a growing field . . . just because we don't have evidence does not mean there are no effects" (Kolata 1999b). Thus, "endocrine modulating chemicals" become "endocrine disrupters" or "gender benders." Manufactured chemicals allegedly cause a decline in human male sperm production, though the latest studies continue not to confirm this (Acacio et al. 2000). Though the argument for a decline in sperm production that began around 1960 is widely disputed, it is interesting that a proponent of the thesis now argues that the cause may have been iodination of salt, which began in the mid-1920s and would have impacted the men who were studied (Crissman et al. 2000). Nevertheless, the message remains that sufficient iodine for brain development is more important than larger testes and a larger sperm count since, if the decline has in fact occurred, it has not given rise to any fertility problems.

Win the name game and you have won most of the public relations battle even when you are otherwise totally lacking in evidence for your fears. The panel created to study the so-called endocrine disrupters felt compelled to call them "hormonally active agents." After winning the name game, the next trick is to ignore the fact that these same human-made chemicals that we ingest in minuscule quantities may be naturally far more abundant in the foods that we eat, such as tofu, which is rich in phytoestrogens. A vegetarian diet by pregnant women has been linked to genital defects in their babies and health problems for growing children (Sanders and Reddy 1994; Sanders 1995; Law 1995; Lovblad et al. 1997; von Schenck, Bender-Gotze, and Koletzko 1997; Hackett, Nathan, and Burgess 1998; NR 1990; North and Golding 2000). The most likely causes are the phytoestrogens or isoflavones in soy products. One research finding has suggested diminution of brain functioning in later life as a result of thrice-weekly tofu consumption (White et al. 2000; Grodstein, Mayeux, and Stampfer 2000).

Stephen Safe estimates that the human diet has forty million times more naturally occurring estrogen-mimic chemicals than the same person gets by being exposed to synthetic hormonelike compounds such as those found in many leading pesticides. Stated differently:

> A mass balance of dietary levels of industrial and natural estrogens . . . indicates that the dietary contribution of estrogenic industrial compounds is 0.0000025% of the daily intake of estrogenic flavonoids in the diet (Safe 1995, 346). Recently, in a comprehensive article that is an update on the latest findings, Safe has added: The role of endocrine disrupters and human disease has not been fully resolved; however, at present the evidence is not compelling (Safe 2000; Safe et al. 2000).

If enough different dangers are conjured, it becomes difficult to deny all of them, since proving a negative is always a problem. The result too often in the absence of proof of danger is to call for more studies. Inherently, there is nothing wrong with that except that it diverts resources from studies that are far more likely to produce results that further the human endeavor.

Human Cost

We sometimes ignore another cost in that lengthy litigation can take a toll upon all the litigants. We have had the specter of years of litigation over the alleged harm done by breast implants that have driven at least one multibillion-dollar firm into bankruptcy. Studies (over twenty by one count) by some of the best scientific research organizations have failed to find any link between silicon breast implants and the alleged health problems. For the women with breast implants involved in the class action suits and for those who were not, the "true health toll of breast implants was the seven years of needless anxiety suffered . . . at the hands of all those lawyers and health 'advocates' who were ostensibly acting on their behalf." Gladwell adds some excellent advice: "tempting though it is to believe otherwise," it is not the business of science to punish the "politically retrograde, nor is it a means of serving retribution to the wicked and irresponsible" (Gladwell 1999, 24).

A recent comprehensive epidemiological study of breast cancer mortality in the UK and USA in the year 2000 found a 25 percent decline in the death rates. Stated differently, "improvements . . . since the 1970s in the way breast cancer is managed must have prevented about 30% or 25% of the breast cancer deaths in middle age that would otherwise have happened in these countries in the year 2000" (Peto et al. 2000a). The use of tamoxifen in treatment has been a vital factor in lifesaving breast cancer treatment (Peto et al. 2000a,b; Lancet 2000e). Unfortunately, the World Health Organization and other health agencies have listed tamoxifen as a carcinogen. Citing tamoxifen as a known carcinogen has led to an outcry by groups who otherwise are quick to label any product as a carcinogen no matter how weak the evidence may be. Tamoxifen may be a carcinogen, but the evidence shows that it prevents the recurrence of breast cancer, and it is thirty times more likely to prevent a death than cause one.

The argument against listing tamoxifen as a known carcinogen is that it might lead to women refusing it, which is a legitimate concern. The fault here lies with those who for years have made careers of alarming us about everything being a carcinogen while making the inane, meaningless, impossible, absolutist demand of zero risk and the public outcry for it. The approach that I am arguing for is one of always balancing risk and benefits and deciding accordingly. When the choice is one harm versus another harm, 30 to 1 odds are compelling to anyone but those who demand zero risk. Bruce Ames showed in his famous 1983 article in *Science,* "Dietary Carcinogens and Anticarcinogens," even foodstuffs that are rightly recognized as anticarcinogenic, such as broccoli or other vegetables and fruits, also contain chemicals in minute amounts that are known carcinogens (Ames 1983; Frary et al. 2000). All plants (edible included) are chemical factories that produce many chemicals, some harmful, some helpful. It is the balance between the carcinogens and anticarcinogens in foods that should govern our actions, not merely the existence of a carcinogen.

Scares about Technological Dangers

Nuclear Disasters

A 1998 study in England found "geographical clusters of childhood cancer cases" and elevated cancer deaths among children "born near industrial sites that use high-temperature furnaces or discharge volatile organic compounds" or near large hospitals (with incinerators). Equally interesting, they found confirming evidence "that nuclear installations and pesticide factories did not present a risk" (BBC 1998b). Thus, it appears that, in terms of danger from cancer, it is safer to raise your children near a nuclear power plant or a pesticide factory than near a large hospital that incinerates its waste. It is also clear that the alleged carcinogenic danger from these nuclear power facilities and pesticide plants has not been verified. Since the study covered the years 1953-1980, presumably this would have included plants producing organochlorines such as DDT. However interpreted, the results are clearly contrary to some of today's most passionately held beliefs about the dangers of radiation and modern chemical pesticides.

Although the 1998 study saw no risk in living near a nuclear installation, there are exceptions. The most dangerous nuclear power station for a child to have lived near was the infamous Chernobyl. A 1986 non-nuclear explosion caused the release of radioactive material, which, by some estimates, killed 15,000 people, caused some illness to 3.5 million people, and did long-term damage to the environment. At the time, there was the reasonable fear that the long-term effect would be a substantial increase in cancer. It did result in 1,800 children developing a seldom fatal, treatable form of thyroid cancer (UNSCEAR 2000; IAEA 2000; IFRCRCS 2000b, chap. 5). The IAEA (2000) maintains that with this exception, there is no scientific evidence of increases in overall cancer incidence or mortality or in nonmalignant disorders that could be related to radiation exposure.

However horrible this tragedy was and continues to be, it was no worse than many other disasters involving large explosions, and it was the result of unauthorized procedures in a type of reactor that did not have the protective structures, mandatory for reactors outside the Soviet bloc, that would have prevented the release of the radioactive material. The average annual radiation dose (expressed in millisievert-mSv) that the local population continued to receive in the year 2000 from the Chernobyl accident was 0.002 mSv. This compares to the "worldwide average annual effective dose" for the natural background of 2.4 mSv, for diagnostic medical examinations of 0.4, from atmospheric nuclear testing of 0.005, and from nuclear power production of 0.002 (UNSCEAR 2000). Radioactivity is dangerous, and nuclear power generating plants should have significant safety factors built in, as they are everywhere (at least outside the former Soviet bloc) required to, which are carefully regulated. Nevertheless, however serious the dangers are concerning nuclear power generation, they are not even remotely close to the level of danger feared by the antinuclear power activists.

Alar and EDB

In the United States and in other advanced industrial countries, there have been scares about technological dangers that have bordered on hysteria. In 1989, in the United States, there was a massive public re-action to apples treated with a chemical called alar. Apples were re-moved from school lunch programs; sales plummeted. One source that wished to have alar banned immediately estimated that a child who drinks an average of 10 ounces of apple juice every day from the first to fifth birthdays would face a cancer risk of from 5 to 50 in 1 million (CR 1989, 291). Looking at the possible total impact of alar-treated ap-ples upon school children, an editorial in *Science* argued that "even in a worst case scenario the probability of cancer among the affected group would change from 25% to 25.05%" (Koshland 1989, 9).

The alar scare was deemed to be a hoax by Joseph D. Rosen, a lead-ing toxicologist, and was included as one of the "health scares that weren't so scary" by Jane Brody, a leading science journalist (Rosen 1990b; Brody 1998). There is a stark contrast between the saturated media coverage of alar and the massive public response compared to the modest coverage and almost no public response to earlier televi-sion reports of known carcinogens, aflatoxins in the corn (maize) and peanut crops because of the stress caused by drought. The disparity in the news coverage and public response gives rise to public policies that fail to weigh comparative risks and act accordingly. "In fact in 1988, the U.S. corn (maize) crop had much higher than usual aflatoxin levels as a result of drought conditions" (Gray and Graham 1997, 187). The fungicide, ethylene dibromide (EDB) was not used on the crop, since EPA (Environmental Protection Agency) had banned it. "It is possible that the public was exposed to a much higher cancer risk through aflatoxin than EDB would have presented" (Gray and Gra-ham 1997, 187).

Problems with Natural Foods

What is natural is routinely not questioned or considered a cause of any of our dietary or health problems. However, our trust can be mis-placed. Organic lettuce was seriously contaminated with a deadly form of *E. coli,* yet it did not draw significant media attention until well af-ter the event. Being organic was not considered in any way the cause of the problem (Belluck and Drew 1998). Did anyone question the as-sumption of the superiority of natural apple juice when many people got sick and one child died from drinking unpasteurized apple juice several years later (Drew and Belluck 1998)?[1]

Despite this experience, unpasteurized orange juice (containing

Salmonella) continued to be sold, threatening human health. Each year in the United States,

> 800,000 to 4 million salmonella infections result in approximately 500 deaths. . . . Juice has been implicated as the vehicle of transmission in at least 15 outbreaks in the United States in this century involving pathogens. . . . The outbreak described in this report is the second and largest salmonella outbreak associated with unpasteurized orange juice (JAMA 1999b; MMWR 1999m; CDHS 2000c).

While improved methods of handling and inspection were able to reverse the general upward trend in foodborne illness in the last years of the 1990s, salmonella continued to increase, reaching a record high from such causes as unpasteurized orange juice, imported mangos, and raw sprouts (MMWR 2000b).

It appears natural contaminants are not to be feared, even if they are highly carcinogenic or associated with other known carcinogens. Bruce Ames finds it "a very funny way of looking at things . . . in the modern Western world," that evil is only a creation of human hands, given the level of health and longevity that we have obtained (Proctor 1995, 140). As I note elsewhere, pasteurization of milk was delayed in its introduction into the United States until the 1920s. One would have thought that some knowledge would have been gained from this delay about the dangers of not using an effective technology for food safety. Both milk and juice are subject to food contamination for which processes like pasteurization can provide us with considerable protection.

Origins of the Organic

Initially, organic agriculture was primarily concerned with using manure instead of artificial fertilizers. An early opponent of the use of artificial fertilizers was the founder of anthroposophy, the Austrian-born German writer Rudolf Steiner, who in the 1920s inspired a new school of farming known as "bio-dynamic agriculture." He advocated self-sufficient farms, which preserved the spirit of the soil and were tilled according to his vision of the life forces and magnetic influences of the cosmos. Artificial fertilizers were rejected because they were man-made (Bramwell 1989, 20; Steiner 1958). In opposing the new labeling rules for organic foods, purist critics are still concerned with synthetic inert substances being used in the production of food.

Writers on "organic agriculture" seem to believe that the urea in fertilizer is somehow different and less potent than the urea in manure. We need a Gertrude Stein to explain to our nature lovers that urea is urea is urea and it is $H_2N \times CO \times NH_2$, whether in fertilizer or manure. Urea was synthesized one hundred years before Steiner's organic advocacy, and since Frederick Wölher's work in the 1820s, organic chemists have had knowledge about its properties that some wish to deny. Organic agriculture enthusiasts view manufactured urea purely as a chemical compound, which it is. But the urea in manure, which is identical to the synthesized compound, is seen as having some additional "vital" or living properties (BBC 2000). Unfortunately, the living properties in organic agriculture may well be the greater endowment of harmful microorganisms that they contain.

Dennis Avery, using the Centers for Disease Control data, finds that "organic foods made up barely 1 percent of the U.S. food supply, but were implicated in at least 8 percent of the confirmed cases" of people infected with the O157:H7 E. coli strain (Avery 1998 a,b, 20; Tauxe 1997b; DiMatteo 1997; Marwick 1997; Stephenson 1997a,b). In other words, "people who eat organic and 'natural' foods are eight times more likely as the rest of the population to be attacked by this deadly new strain" of bacteria. They are at greater risk of being attacked by a relatively new, more virulent strain of the infamous *Salmonella* bacteria (Avery 1998b, 19). Compounding the tragedy of children dying from *E. coli* is the emerging evidence that for those who survive, there is a seventeenfold increased risk of a kidney disorder, "Hemolytic-Uremic Syndrome, after antibiotic treatment of *Escherichia coli* O157:H7 infections" (Wong et al. 2000).

Natural Toxins

According to data from the U.S. Food and Drug Administration, crops grown organically have higher rates of infestation by natural toxins, such as the aflatoxins from the mold *Aspergillus flavus* (Avery 1998a, 22). Despite this evidence, the media not only continue to be concerned about pesticides in food but also warn of a government cover-up of pesticide dangers and the potential benefits of organic agriculture. Dennis Avery (1998b, 22) asks: "What is more dangerous, pesticides or horse manure?" He answers,

Researchers are still looking for the first death from pesticide residues, fifty years after DDT was introduced and thirty years after its use was banned in the United States, but manure is claiming lives almost daily through bacterial contamination of organic food.

At about the same time as the U.S. alar scare, in the United Kingdom there were two outbreaks of serious natural food contamination, salmonella, listeria, and clostridium botulism. There was once again in 1999 a problem with listeria and a very serious outbreak of the O157:H7 *E. coli* strain, apparently from unpasteurized milk and cheese. These traditional food contaminants have become more difficult to control because of the consumer's increasing aversion to preservatives and what they consider to be other food contaminants.

A 1997 report by the World Health Organization found that the incidences of food poisoning are "seriously underestimated." Most occur in developing countries and all appear to be "natural" (Tauxe 1991). Foodborne pathogens are the sixth leading cause of death in the world, claiming about 2.5 million lives each year. The fact that few people in developed countries today die from botulism is also a result of modern medical developments, many of which are opposed by our alleged nature lovers (Glynn 1989, 671–72).

The honest attempt by consumers to obtain healthier food for themselves and their families often becomes an example of what has been called the strategy of the riskier alternative as the risks from the consumption of "organic" food are either ignored or denied. Any hint of carcinogenicity from feeding large doses of a chemical to experimental animals, mostly to rodents, is taken as definitive and final while clearly understood risks from food contamination by microorganisms is simply denied. Fortunately, a "philosophy is taking hold" that recognizes that high-dose studies for toxicity are not always the best indicators of a threat to human health. New guidelines "require regulators to incorporate into risk assessments such factors as how a chemical's structural features might affect its toxicity and how potential poisons are absorbed, metabolized, and distributed in the body" (Stone 1995, 356).

Bruce Ames argues not only that fungus infestation of plants is dangerous in and of itself but also that such infestation causes plants to "produce very much larger amounts of their natural toxins," many of which are likely carcinogens (Ames, Magaw, and Gold 1990, 78, 80; Abelson 1994b). According to Ames, "we are ingesting in our diet at least 10,000 times more by weight of natural pesticides than of man-made residues" (Ames, Magaw, and Gold 1990, 75; Ames, Profet, and Gold 1990a,b; Ames and Gold 1991). He estimates that the light-activated carcinogens in celery can "increase 100-fold when plants are damaged by mold and in fact can cause an occupational disease in celery-pickers and in produce checkers at supermarkets" (Ames, Magaw, and Gold 1990, 80; French 1990, 15–16).

Ames also argues that very low doses of some chemicals may actually be beneficial in helping our immune system to later withstand larger doses (Ames 1992; Calabrese 1994). This finding corresponds with a general theory, known as the hygiene hypothesis, that modern public health measures such as immunization have reduced our exposure to disease vectors, therefore not sufficiently challenging our immune system (apart from the immunization itself), and thus preventing it from becoming more fully developed. This has been used to explain the rise of asthma and other atopic diseases (Strachan 1989; NIEHS 1997). The thesis has been tested over the years with varied results but remains sufficiently viable as a theory that it continues to be tested (Paunio et al. 2000; Matricardi et al. 2000; Voelker 2000a; Gern and Weiss 2000). Our environment may be too aseptic (Ernst and Cormier 2000; Levy 2000; ASM 2000a,b). Nevertheless, pesticides and outdoor air pollution continue to be blamed for the increase in asthma. A more recent study in the UK finds a decrease in the incidence of asthma and rejects the thesis of "chemicals" being a cause of asthma (Fleming et al. 2000a,b). However, a recent panel study by the Institute of Medicine, National Academy of Sciences, found household causes of asthma to be more likely the result of natural factors, such as mites and cockroaches (IOM 2000a; NAS 2000a; Lancet 2000d). As the report recommends pest extermination, one might reach the seemingly perverse conclusion that more spraying of pesticides would produce a healthier house. Research has shown that the manufactured chemical household disinfectants are more effective in "killing disease-causing organisms, eliminating 99.9 percent of the microbes," than "natural products" such as "vinegar and baking soda" (Williamson 2000).

Fears about genetically altered microorganisms abound, yet the microorganism that is causing the most serious condition, AIDS, was identified only as a result of the advances in biological research for which so many concerns are raised. We have the paradoxical situation where there is a very reasonable hypothesis that our environment may be too aseptic while there is also concern that the "health" food lifestyle of organic foods, unpasteurized juices, and raw sprouts may be encouraging the spread of foodborne pathogens that were largely under control. They both could be correct, which means that those with immune systems that have not been sufficiently challenged are in even greater danger when encountering foodborne pathogens or newly emerging ones.

Whatever the case may be, it is abundantly clear that never in the history of humankind have we had cleaner food and drink. Our hunter-gatherer ancestors were sickened by pathogens infecting the wild animals they killed, and the water of even purest stream was often full of pathogens from animal feces or other contaminants. The situation may have become worse with agriculture, urbanization, and more densely packed humanity until modern science and technology provided the means for clean food and drink.

The organic agriculture movement arose in opposition to the twentieth-century developments in agriculture that have created our modern food system with all its achievements and all its potential to complete the task of abolishing hunger and malnutrition. Organic agriculture as a movement is today driven by technophobia and sustained by opposition to continued technological progress in agriculture and society. If the practice of organic agriculture were just an alternative to modern agriculture, few of us would take the trouble of criticizing it, as it would constitute another way from which we all might learn something. As I attempt to show throughout this book, it is the active attempt of the organic practitioners to justify their existence by undermining vital technological change that has become the source of a growing divide and ideological conflict.

However severe the economic, technoscientific, and environmental difficulties may be to an expansion in food supply, overcoming the barriers of technophobia is as great an obstacle to change as any other. Realizing the promise of food sufficiency and food security for everyone requires continued technological progress and a strong commitment—financial, ethical, moral, scientific, and technological—from all to make it happen.

Note

1. Symptoms of *E. coli* infection include acute abdominal pain and blood in the stool. Young children are the most prone to develop complications of *E. coli* infection, which include kidney failure and other potentially fatal problems (Priorities 1996, 5; MMWR 1997).

CHAPTER 4

Technophobia, Natural Foods, Organic Agriculture, Energy, Meat, and Metamorphosis

Vegetarianism neither is the "original human diet" nor is more natural than any other diet. Not only have humans been meat eaters as long as they have been human, but our closest relatives, the chimpanzees, are also meat eaters. From about 200,000 years ago onward, the hominid brain has grown larger, requiring increasing amounts of energy to maintain it, but the gut has grown smaller, reducing the ability to extract energy from low-density foods such as the high-fiber vegetable matter eaten as the primary food by other primates like gorillas. "Since the time of *Homo erectus* and the gradual transition into modern humans some 200,000 years ago, human brain size exploded" (Stanford 1999, 175; see also, Milton 1993, 91), leading to the creation of "an apparent anomaly in that we have a species with an unusually large brain, an organ regarded as an energetically expensive one, and yet at the same time this species has a gut that appears small for its body size" (Milton 1988, 301). The overall smallness of the human gut is in terms of the large intestine or hindgut, since the human small intestine is relatively large compared to other primates. Milton notes,

Brain tissue is metabolically expensive, requiring large supplies of oxygen and glucose. Furthermore, the brain's demand for these products is constant and unrelenting, regardless of the mental or physical state of the organism (Milton 1988, 299).[1]

Hunting and Gathering

The foods of our ancestors and their methods of acquiring them made them what they became. There is a fascinating body of anthropological literature on how the search for gathering and consuming high-nutrient and energy-dense foods (such as fruits and nuts) provided humans an evolutionary advantage in the development of memory, communication, and social skills, and how hunting furthered these and other intellectual abilities necessary for the later development of civilization. Vegetable matter was relatively ubiquitous, while fruits and nuts were often widely scattered and seasonal, and, except for scavenging, meat had to be pursued and caught (Milton 1987, 107; Milton 1981; Milton 1999). Thus, Milton says that

> Foraging behavior of early humans was probably marked by increased complexity in communication skills, social skills, and technological skills . . . and all of these behaviors are assumed to probably be related to greater mental complexity in our lineage (Milton 1988, 302; also Milton 1987, 106; and Milton 1993, 88–89).

The evolution of hunting was vital for the development of human intelligence, particularly the "transition from pursuit hunting to the sort of ambush hunting that requires a greater degree of planning and may indicate higher cognitive development" (Stanford 1999, 134). Stanford (1999) says, "Humans living 20,000 years ago had also learned to plan their hunts in advance, and to drive game through narrow bottlenecks in the landscape where they could be more effectively ambushed" (Stanford 1999, 135).

Hunting involved foresight, planning, the accumulation of knowledge (of the prey) and anticipation and thinking-in-time. In other words, hunting required those abilities of thought that we used in the opening chapters to define being human. They may also have known "the annual migratory cycles of the hoofed game and followed them, or planned their hunting calendar around the much-anticipated movements of game through the area. Hunting by modern men thus became routine, strategic, well coordinated rather than opportunistic" (Stanford 1999, Stanford 1999, 135). Stanford adds that "the roots of intelligence may lie in a combination of ecological complexities, the value of foresight in making and using tools, and the value of being socially intelligent" (Stanford 1999, 176).

When meat eating became essential for human survival, then a vegetarian diet could in no way be considered natural. Ironically, if the word "natural" has any meaning in the context of such a culturally saturated endeavor as eating, then vegetarianism would be the least natural in that our hominid ancestors could not have survived on it. The modern vegetarian lifestyle is possible because through domestication of fruits and nuts, domestication and milling of grains, cooking etc., and transportation, refrigeration, and affluence, a vegetarian can regularly acquire the necessary nutrients (except vitamin B_{12}, which comes only from animal products) in sufficiently refined and palatable forms to survive. In other words, contemporary vegetarianism is as unnatural as any other activity in modern life. Modern food production has made the obtaining of adequate nutrition even easier for the lacto/ovo vegetarians.

The Natural Diet

If not vegetarian, what is a natural diet? If "natural diet" is meant to imply any possible tendencies that we may carry as an evolutionary heritage and that may manifest themselves through cultural preferences, then we would have to look at the dietary preferences of earlier ancestors where the cultural overlay would be less imperative, who would be less affected by cultural setting. William Hamilton (1987, 125) observed that the primate preferences patterns when alternatives are available were:

(1) Least preferred: leafy material, including grasses;
(2) More preferred: rhizomes, fresh grass seeds, fruits, and flowers;
(3) Most preferred: animal matter, including but not limited to insects.[2] (Hamilton 1987, 125, see also Cohen 1987, 266)

This is in line with Katharine Milton's (1988) observation that "the large size of the human small intestine appears characteristic of all human groups, regardless of diet, and indicates that

(1) all modern humans eat a relatively high quality diet;
(2) the feature is under genetic control;

(3) it is derived from the common ancestor of all extant human beings. It further suggests that this common ancestor ate a diet rich in nutrients and energy dense (Milton 1988,301).

Increasingly, hominids did not have sufficient digestive capacity to subsist on less energy-dense foods without including high-quality, more energy-dense foods. This also meant that high-fiber, low-quality food had to be passed by the gut more quickly in order for there to be room for the high-quality foods (Milton 1993, 93; see also, Milton and Demment 1988). It is likely that humans developed a craving for high-energy, nutrient-dense foods because of the biological need for them, the difficulty of obtaining them, and because early human diets were deficient in them.

A high-fiber diet today is high in fiber relative to our contemporary diet and not in any evolutionary sense; one of its virtues is as an aid in passing food quickly through the gut, minimizing the gut's absorption of possible toxins. Similarly, "whole grain" is also a relative term. Our ancestors, in consuming those energy-dense seeds, had trouble digesting them and extracting the nutrients from them. The evolution of wheat from its wild ancestors, for example, can be traced through undigested seeds found in fossilized human excrement, called coprolites. The Indians of Baja California's "practice, termed 'second harvest,' was the custom of removing seeds from human feces, grinding them up, roasting and eating them" (Newcomb 1986, 41). Humans could only efficiently and consistently obtain the nutrient from the seed after the advent of milling and cooking.

Again, from Hamilton (1987) we learn that "overconsumption of meat and sugar by some human populations may be based *in part* upon dietary predilections of our omnivorous ancestors" (Hamilton, 1987, 118). Tasting "sweet" in nature would be a good proxy for caloric density to an animal, as would fat.

> If we assume that human ancestors possessed in general form the dietary predilections of contemporary omnivorous primates, it follows that with improved hunting and other procurement abilities, the proportion of animal matter in human diets could and would increase (Hamilton 1987, 121).

The food preferences that we are born with as biological beings were evolved under conditions of scarcity of meat and, to a lesser extent, of fruits (that which tasted sweet). That we may consume animal fats and sugars to excess today does not negate the need for these items in our diet in more reasonable proportions. Hamilton (1987) says, "The quantity of animal matter and products in the diets of some contemporary people can be understood in terms of a nearly unlimited availability of animal matter superimposed upon food category preferences in the past" (Hamilton 1987,121).

Hamilton offers the hypothesis that "some humans are led by inherited biases to choose a diet that our ancestors sought but could not have obtained for any practical interval" (Hamilton 1987, 121). "Sugar overconsumption can be explained in the same way. Fruits are the next choice after animal matter for omnivorous primates. However ripe and sweet fruit is only seasonally available" for primates but not for modern humans (Hamilton, 1987, 121–122). It is quite possible, then, that the current diet of affluent Americans which is too rich in fats and sugars is the result of "natural" cravings manifested through our cultural food preferences. This may explain why it is so difficult for many of us to limit our intake of these, even though we know better. If we succeed in appropriately diversifying our diet, it is not because it is more "natural," but because modern knowledge so guided us.

Technology, the "Organic," and "Natural" Foods

The Organic Food Production Act of 1990 mandated the creation of standards for "organic" food by the U.S. Department of Agriculture. No pesticides or synthetic fertilizers and hormones would be used to grow or process it. One of the prevalent myths about organic foods is that they derive nutritive superiority from using manure, not artificial fertilizers. There is simply no verified or verifiable difference in the taste or nutritional quality between "organically" grown food and those using modern methods of fertilization. In my judgment, to the believers, organic produce tastes better because it has to; otherwise their beliefs are undermined. I think those who are absolutely certain that the "organic" product tastes better are dealing more with imagination than fact. Presumably they have to get something back for the higher prices they pay.

Organic nutrient must be broken down by the soil before it can be absorbed by the plant, which then proceeds to use the nutrient as components to manufacture the complex molecules for growth and development. The information about what plants can absorb is in most any basic text on agriculture. Those who would have us live in "harmony with nature" fail to realize that this is inherently impossible. In evolving with our technology, we survived outside our original "natural" environment by transforming or disturbing it to make it suitable for our habitation and for increasing densities of population (DeGregori and Thompson, 1991). Wood (1998) argues that there is "sound ecological support for the concept of 'farming as disturbance'" (Wood 1998, 373, quoted in Bazzaz 1987, 248).[3] This does not mean that we are not still part of nature but that we have transformed our role in nature.

The Harmonious Natural World

Many of the metaphors for a harmonious natural world are derived from developments in technology (clocks) and idealization of other mechanical creations (Botkin 1990, 101–106). The nineteenth-century romantics added various concepts of "aliveness" (Henri Bergson's *élan vital*) and holism to the more mechanistic models of the natural world (Drury 1998, 17).

Nature was viewed as having an "unseen hand" (much like the "invisible hand" of Adam Smith); as a concept of the economy of nature, energy flows through the system. Balance of nature, equilibrium, succession, and biotic community emerged (Drury 1998, 17–22). Many of these ideas are teleological, as an ecosystem undisturbed by mankind moves inexorably, following species succession in a self-organizing process, each species preparing the way for its successor, to a foreordained ecological climax. The outcome is a "particular, preferred configuration of species characterized by a variety of 'good things': maximum productivity, diversity, efficiency, large biomass, nutrient cycling, stability, 'information content,' and so on" (Drury 1998, 184).

There is more than a superficial similarity to maximization theory in economics, which is ironic, since those who follow the precepts of ecological economics or allied views reject the "invisible hand" of Adam Smith and maximization models in economics. Ensuing from these deterministic ecological equilibrium models is a "laissez-faire" conservation/preservationist policy that would exclude any human activity for being disruptive of nature's balance. In practice, one really does not know whether removing a tract of land from active human use will in fact save any species from extinction (Drury 1998, 187). Drury says,

"Much ecological theory includes the assumption that natural ecological change is progressive, that an end point is an equilibrium community that is the fullest development of local ecological community" (Drury 1998, 210).

Drury likens this type of ecological theory to the "primary principles underlying the Great Chain of Being," namely "plenitude, continuity and gradation" (Drury 1998, 21). He asks the question as to why "we continue to imagine that self-organizing principles and holistic benefits exist in natural 'communities.'" Basically, it is emotionally comforting to do so, as we "put our confidence in supernatural (self-organizing) principles rather than address awkward realities" (Drury 1998, 24). Unfortunately, what is called the "ecological movement," though it has little relation to ecology as a scientific inquiry, is based on a "model of stable communities which has been largely discredited during the last two decades." Drury argues that conservation (and agricultural) practices built on "outdated ecological models will eventually backfire" (Drury 1998, 185, for the diversity-stability debate, see, Mc-Cann 2000). Kevin McCann (2000) provides an excellent overview of the current diversity/stability debate in ecological theory.

Agricultural Disruption

Agriculture is a disruption of the environment and almost always involves defending the crop from competitors and predators, namely animals, insects, or micro-organisms. That does not mean the habitat was "peaceful and harmonious" before human agriculture, rather, that agriculture has substantially transformed the environment in and around it. Grains and many other crops required an initial clearing of the land, thereby removing existing competitive plants and animals, and providing some form of tillage to aerate or otherwise prepare the soil for domesticated plants. In most instances, it required amending the soil either with silt from floods, burning plant residue, or, as in more recent times, adding the basic NPK (nitrogen, phosphorus, potassium) fertilizer, trace elements, or lime to reduce the acidity of the soil. Most of the world's current agricultural lands were not suitable for agriculture without modification by humans (DeGregori 1985, 97). The cultivators must provide adequate water if the rainfall is insufficient. Good conservation practices normally favor as much recycling of nutrients as possible, but you can't recycle what is not there, so initially, most soils had to be amended.

The more successful agriculture is, the more it concentrates high-quality nutrients. With the exception of hydroponics, greenhouses, or other high-tech, capital-intensive methods, most agriculture occurs in the unprotected outdoors. What is nutrient for humans is also nutrient for the life processes of insects, birds, rodents, other animals, bacteria, fungi, and viruses. Some plants also will seek to recolonize the land, the seeds of which are probably in the fields before planting or, prior to modern agriculture, were inadvertently planted along with the crop. Improvements made in the land for the crop also render it more "attractive" for many other plants. Such competitors do not respect the "property rights" of humans any more than humans respect the rights of prior plant and animal occupants of the land.

Though the energy comes from the sun, and some of the constituent elements of the nutrients are derived from the air (and are "naturally" replenished), other nutrients arise from the soil. If food is grown in one location and eaten in another, then, in effect, the soil is being mined, and soil nutrients are carried to another location. As a general rule, the larger the crop, the more nutrients are taken from the soil. Add more nutrients per unit of land with high-yielding varieties and you get a larger output per unit of land, or add the same amount of nutrients to a larger area and you get the same total (but smaller) output per unit of land.

The greater nutrient requirement of the "green revolution" crops per hectare or acre is *purely* a function of their larger output, since most are more efficient, producing a higher yield per unit of input.

Reduced to basics, one can see that there is no magic to agriculture. If you grow it, you must protect it. If it is grown in one place and eaten elsewhere, the cultivators must replace soil nutrients unless they are "naturally" replenished by silt from a flood, volcanic ash, etc. or some other method. Various agricultural methods to address these conditions have been used, and others are being proffered.

Crop Protection and Nutrient Renewal

It is fantasy, however, to suggest that we can grow crops and feed the world's population without some form of crop protection and soil nutrient renewal. Crops have been and are being bred to be more resistant to various forms of stress (soil acidity, water, other input deficiency or excess, disease). Disease resistance, whether "natural" or enhanced through selective breeding (or more modern means), generally involves the manufacture of toxins, which in large enough doses are also toxic to humans. Biological means of crop protection (now generally lumped under the rubric of IPM (integrated pest management), such as various schemes of crop rotation and/or introducing insects that prey on crop-eating insects, have had a mixed history of agricultural success and failures but have increasingly become part of the arsenal of crop defense. Attempts by entomologists to use predator insects to control crop pests in agriculture in the United States began in the 1870s or earlier. The first significant success was in the 1890s with the importation of the vedalia beetle (*Novius Rodolia cardinalis*), "an insect that makes all its meals on the eggs and larvae of the cottony-cushion scale" that was devastating the citrus crop in California (Whorton 1974, 13).

Innumerable substances have been used to protect crops. A major impetus for the rise of the modern consumer movement in the United States in the 1920s and 1930s was concern over the use of lead arsenate as a pesticide (Paehlke 1989, 24; and Whorton 1974, 24, 189–194). Other poisons used for crop and livestock protection before modern chemical pesticides include: the alkaloid nicotine, copper acetoarsenite, potassium 4,6-dinitro-o-cresylate, lime sulfur spray, hydrogen cyanide, sodium arsenite, and potassium antimonyl tartate (Metcalf 1980, 220; and Metcalf 1986, 253, and 259). A solution of inorganic salts, called Bordeaux mixture, is still used in "organic agriculture," even though the copper (often in the form of copper oxychloride or copper sulphate and lime) that is in it is "toxic at the levels used." The "justification" for its use appears to be that "it was hallowed by time" (MacKerron et al. 2000, 68). "Organic" agriculture also uses sulfur, methaldehyde, and a favorite of many, derris or rotenone ($CH_{23}H_{22}O_6$), which was originally derived from the roots of a vine from tropical Asia.

Consumer Reports Testing

After hearing a radio report that *Consumer Reports* did not test for the residue of sulfur and copper salts in "organic" foods, and having been a subscriber off and on for over forty years, I asked for clarification, hoping that the report was false. They not only admitted that they did not test for what they deem to be "natural pesticides" but also claimed that it would be difficult to test for them. They said "there is little basis in terms of actual data to suggest that they, in fact, pose any risk," so they "settled for comparing residues of conventional insecticides, fungicides and herbicides" (CU, 2000). To say the least, it is an odd procedure for a consumer testing organization to make a comparison between conventional and "organic" food by testing only for the toxins used by conventional food while not even indicating to the unsuspecting reader that there may be some unanswered questions about the safety of the toxins used by the other, since *Consumer Reports* concededed that a "few might, in theory" pose a "significant risk" (CU, 2000).[4]

I consulted toxicologists who disagreed with both the letter's claim that the toxins do not "in fact, pose any risks" and the claim that testing for the so-called "natural pesticides" would present any major difficulty for a consumer testing organization. Though *Consumer Reports* proudly and rightly does not take any advertising at all, its current behavior, at variance with the earlier reports, might in this instance have a reason. They may be influenced by both the fact that they accept money from an activist environmental group, the Natural Resources Defense Counsel (which was responsible for, among other things, the alar hoax) and by the fact that the letterhead of an anti-pesticide, anti-genetically modified food activist group, the Pesticide Action Network, has a member on its council who is identified as being with Consumers Union. This is hardly the basis for objective, bias-free consumer information based on scientific product testing and raises the question as to whether Consumers Union is yet another once-reputable organization that has been taken over by zealots.

Pesticides

Prior to DDT, authorities in the United States tried to control mosquito breeding for malaria protection "by casting kerosene and/or Paris Green (a powdered arsenical formula) on the waters without much success" (Desowitz, 1997, 205). Paris Green is copper aceto arsenite, a combination of copper acetate—$Cu(C_2H_2O_2)$—and arsenic trioxide—$3Cu(AsO_2)$—and was even more widely used as a pesticide in agriculture along with other arsenic compounds such as London Purple, a byproduct of the dye industry. Modern antibiotics, beginning with penicillin, also replaced arsenic compounds in the treatment of syphilis (Desowitz 1991, 126–127).

It is interesting to note that when DDT and other organochlorines were first used, the earlier pesticides were often referred to as the "nonorganic pesticides." Whatever may or may not be the problems of modern chemical pesticides, they are definitely benign to the worker and consumer compared to the many substances used throughout human history. And if some scientists are correct, they are also far less toxic than not using pesticides at all.

Rachel Carson, in her highly influential book, *Silent Spring* (seen by friends and foes alike, as the founding treatise of the modern environmental movement), compares the "endless stream of synthetic insecticides" to the "simpler inorganic insecticides" of pre–World War II days. The latter were "derived from naturally-occurring minerals and plant products, compounds of arsenic, copper, lead, manganese, zinc and other minerals, pyrethrum from dried leaves of chrysanthemums, nicotine sulfate from some of the relatives of tobacco." The "new synthetic insecticides" are set apart from the earlier "naturally occurring" ones by their "enormous biological potency" and their "immense power not merely to poison but to enter the most vital processes of the body and change them in sinister and often deadly ways" (Carson 1962, 5). The reader of *Silent Spring* might take this to be an extremely naive endorsement of the earlier pesticides. Carson does go on to indicate throughout the book the dangers of arsenic and lead. But the above statement, even in context, either gives the farmer no choices in protecting his or her crop or would favor using the more "natural" pesticides, as they are allegedly less potent and less insidious in the dangers that they pose.

Fertilizers

Broken down into components, manure and artificial fertilizers are indistinguishable to the plant. There are two important differences. Artificial fertilizers can be composed of the right combination of nutrients for the soil in order for the crop to be grown. The nutrients of manures vary and do not always have the correct proportions to meet crop's needs. Manure is generally considered better for the soil structure, though it poses other problems in that it can have toxic chemicals or a high salt content and may harbor harmful bacteria, insects, worms, and other pests.

Even if manure were in fact superior, it is no longer possible to use it to provide soil nutrients to raise crops and feed the world's population. One estimate is that the world has only about "20 percent of the organic plant nutrients needed to sustain current food output" without even considering the three-fold expansion of farm output required in the twenty-first century (Avery, 1999, 7). In the late 1980s, "about 4.4 billion tons of composted organic animal manure (1.5% nitrogen on a dry-weight basis)" were needed "to produce the equivalent of the 65 million tons of chemical nitrogen" used at that time. As the use of chemical nitrogen increases, so obviously would the volume of manure needed to replace it. "To produce it would require a three- to four-fold increase in world animal production, necessitating, in turn, huge increases in the output of feed grains and pasturage. Furthermore, the transportation costs associated with distributing these fertilizers would be prohibitive" (Borlaug and Dowswell 1988). The high cost of transporting manure from feed lots where cattle are concentrated has made manure become a major pollutant, indicating that even at the current high price for manure, it is too costly to be transported to crop land except in the immediate vicinity.

Currently, production of chemical nitrogen remains between 80 and 90 million tons, supplying better than twice as much of the world's nitrogen needs for agriculture as does manure (about 35 million tons of nitrogen) and about 40 to 50 percent of all the nitrogen needs of the world's crops (Smil 1991, 573, 575, 577 and 580; and Smil 1993, 69–70). "By the late 1980s, this also meant that these [synthetic] fertilizers provided an identical share [40–50 percent] of all nutrients incorporated in the annual global crop harvest" (Smil 1991, 577). Given that "about three-quarters of all nitrogen in proteins comes from arable land, and if synthetic fertilizers provide about half of all nitrogen in harvested crops, then at least every third person, and more likely two out of every five, gets the protein in the currently prevailing diets" thanks to synthetic fertilizers (Smil 1991, 582; see also Emsley 1994, 211).

Not only do we need manufactured fertilizer to maintain current output, but also there will be an ever greater need for more manufactured fertilizer if food supply is to keep up with population growth. The success that using manufactured fertilizer has had by increasing productivity in food production worldwide demonstrates that "there is still an enormous untapped potential for increasing yields" (Oerke et al. 1994, 30). Currently, the lowest application rates for fertilizer are in areas such as Africa where higher food production is desperately needed to increase per capita food consumption of expanding populations. Smil (1997, 76–81) recognizes that the massive use of nitrogen fertilizers is causing serious environmental problems, requiring more effective, efficient use of them, as well as a variety of other innovative agricultural practices. (Smil 1997, 76-81). However, if farmers attempt to return to "purely organic farming, they would quickly find that traditional practices could not feed today's population" (Smil 1997, 81).

Cost of Eliminating Chemical Pesticides

Pest management in modern agriculture has been largely a function of the use of chemical pesticides. IPM, breeding plants for pest resistance, and various types of cropping patterns are increasingly being used, but agriculture still remains dependent on a variety of chemicals, though the dosage level has been greatly reduced. Most serious studies of the use of chemicals in agriculture find that not using chemical pesticides would greatly reduce crop production. Some studies have estimates of global crop losses as high as 70 percent if the use of chemical pesticides suddenly ceased (Oerke et al., 1994). Knutson has collaborated on a number of studies and published numerous monographs on potential losses for specific crops as well as overall surveys of the cost of eliminating chemicals in agriculture. In general, they estimate a 27 percent reduction in output of corn (maize) in the United States without the use of herbicides and fungicides (except for coating the seed) and a 53 percent reduction using no chemicals at all, including fertilizer. For wheat production, the estimates are declines of 24 percent with no pesticides and 38 percent with no chemicals, while peanut production would plunge 78 percent (Knutson et al. 1990, 5 and 6). Another researcher argues that there would have been a 100 percent loss of production for some crops in the United States without the use of pesticides (Farah 1994, 1).

Implications of Organic Agriculture

Many who want a return to "organic" agriculture would have us cut back, if not eliminate, the consumption of meat, and many lacto/ovo vegetarians protest the raising, transporting and slaughtering of calves, though if cows are to produce milk, calves are necessary (Groves 1995). Simply stated, they wish us to use more manure while simultaneously having us produce less of it. And many "organic" purists oppose using sewage sludge because it contains industrial waste in addition to human excrement. If in fact the purists are correct that using sewage sludge for growing crops poses dangers to consumers' health, then it is one more argument for the necessity of using synthetic fertilizers in order to be able to sustainably raise crops and feed humans. If we "mine" the soil for nutrients to raise crops, transport the nutrients to where the humans are who will consume the crops, and dispose of the waste in ways that do not allow its return to agricultural land in the foreseeable future, then it is clear that we have to synthesize packages of nutrients (fertilizer) and return them to the soil.

Every environmental and health sin imaginable is attributed to the raising of cattle and the consumption of meat, particularly beef. If we were to increase manure production by keeping more cattle, there would be costs. The environmental costs of bringing more land into cultivation for grain crops, converting more land to pasture, and hauling several billion tons of manure would undoubtedly be far greater than those environmental costs that result from the manufacture and transportation of chemical fertilizers.

Food Faddism and the "Counter Cuisine"

The label "organic" on food products makes consumers think they are getting healthier food and some are willing to pay higher prices. They can sometimes get more than they expect, however.

Even regular organic food costs more, and any substantive change in that direction would raise the price of food, adversely affecting the poor (NRC 1989, 23 and 55). Some pay up to three times as much to buy organic produce, even though it has now become widely available. *Consumer Reports* (1998a) finds that organic foods cost 57 percent more on average (CR 1998a, 16, see Gray and Graham 1997, 190).

The organic movement has given us the principles of the "counter cuisine," which are based on "inversions of established beliefs." Inverting established notions of spoilage, the counter cuisine equated preservatives with contamination and microbes with health. "Decay was healthy, the more biodegradable the better. . . . Natural warmth was welcomed as an ally of natural decomposition." Warmth is "healthy," provided a variety of micro-organisms whose growth would be facilitated are bearers of good health (Belasco 1989). Problems can occur because of other microorganisms: "Perishable foods contain nutrients that pathogenic microorganisms require to reproduce. Bacteria such as *Salmonella* sp., *Clostridium* sp., and *Staphylococcus* sp. can multiply quickly to sufficient numbers to cause illness. Prompt refrigeration slows bacterial growth and keeps food fresh and edible". (MMWR 1999h, see also; Bruhn 1997). To the counter-culture and its cuisine, energy-wasting refrigeration is to be "scorned" (Belasco 1989, 40).

Belasco is consistent as he dismisses concerns about botulism, salmonella, and dysentery that have plagued humanity and still devastate the poor throughout the developing nations. (Belasco 1989, 119). The "counter cuisine" is likewise scornful of any concern for natural toxins in foods and the aflatoxins from smut or rust that infest grains (Belasco 1989, 120). The lack of concern about microorganisms is indeed strange. Garn (1994) says, "Effects of rust-contaminated grains can also be variable and cumulative, increasing over a season and with particular impact on pregnant women and their fetuses not evidenced until many months after their ingestion" (Garn 1994, 91). Conveniently ignored also are the various contaminants of sludge and other "organic" fertilizers, including microbes and metals. Referring to the deadly new strains of *E. coli* and salmonella, Avery (1998b) and Ekperigin (2000) argue that animal manure is "the biggest reservoir of these nasty bacteria that are afflicting and killing so many people." We are often told that the heat from composting will kill the microbes. A temperature of 160 degrees Fahrenheit is required to kill harmful microorganisms. Few organic farmers actually use a thermometer to test the temperature. Further, studies indicate that "*E. coli* 0157 bacteria can live at least seventy days in a compost pile" and therefore it would take an "extended period at 160-degree heat to kill it" (Avery 1998b, 21). Even if the compost reached this temperature, organic agriculture advocates assume that organisms can evolve to become resistant to pesticides but are unable to evolve to survive elevated temperatures. Even recycling household wastes for use in agriculture can be harmful to the health of the household members unless special precautions are taken (Wouters et al. 2000).

Organic farming increases risks of "fecal contamination not only of foodstuffs but also of waterways, food poisoning, high levels of natural toxins (e.g., aflatoxins) and allergens, contamination by copper- and sulfur (contaminated with lead)-containing fungicides" (Hillman 2000, 28). This has resulted in "production of blemished, diseased and irregular produce of low consumer and food processor acceptability, low productivity, and creation of reservoirs of pests and diseases" (Hillman 2000, 28).

Mortality in the Twentieth Century

With the exception of the Spanish influenza pandemic of 1918, there was a steady decline in mortality or morbidity from infectious diseases throughout the twentieth century until the 1980s. HIV (AIDS) was clearly the main force for this increase in infectious disease rates, but foodborne pathogens have also contributed and are becoming more important as the deaths from AIDS have leveled off; it is no longer one of the ten leading causes of death in the United States. An increase in foodborne diseases is attributable in large part to the "organic" food movement and the larger lifestyle with which it is associated. In the United Kingdom, food poisoning deaths ("overwhelmingly from bacterial infections") doubled in the last decade of the twentieth century at a time when the consumption of "organic food" increased tenfold. There is strong evidence to believe that there is a causal link between the increased consumption of "organic food" and the increase in food poisoning (Bate 2000, for the counter argument, see SANR 2000). Needless to say, the Soil Association (UK), one of the oldest organic growers' associations, has an exactly counter view (SANR 2000).

Impact of Foodborne Disease

Researchers at the Centers for Disease Control and Prevention attempted to "better quantify the impact of foodborne diseases on health in the United States" by compiling and analyzing information from multiple surveillance systems." They estimated that "foodborne diseases cause approximately 76 million illnesses, 325,000 hospitalizations, and 5,000 deaths in the United States each year. Known pathogens account for an estimated 14 million illnesses, 60,000 hospitalizations, and 1,800 deaths" (Mead et al. 1999).

Though "known pathogens" account for only a minority of foodborne diseases, three pathogens stand out as major causes. "Three pathogens, *Salmonella, Listeria,* and *Toxoplasma,* are responsible for 1,500 deaths each year, more than 75% of those caused by known pathogens, while unknown agents account for the remaining 62 million illnesses, 265,000 hospitalizations, and 3,200 deaths" (Mead et al. 1999).

The aflatoxins secreted by the fungus *Aspergillus flavus,* which infects maize (corn), rye, and ground nuts (peanuts), are one hundred times more carcinogenic than PCBs were once thought to be, have brought untold misery to humans throughout history, and still are a source of misery and death in poorer countries. Aflatoxin is still an important liver carcinogen (Doll and Peto 1981, 1228; and Henry et al. 1999). The same can be said for the toxins, ergonovine, ergotamine, and other alkaloids of ergot, a *Fusarium* mold or fungi from the genus *Clavicepts* (ascomycetes), which, among many other things, are abortifacients, possibly hallucinogens, and immunosuppressants, weakening the resistance to infectious disease.

Before 1940, anemia, beriberi, pellagra, and ariboflavinoses were common in the United States, causing illness and death. Now "it is difficult to find a case for study" of vitamin deficiency diseases that numbered in the hundreds of thousands of cases in the 1930s. "The number of deaths due to pellagra in 1966 was 1.1 percent of the figure for 1941. In 1921 three-fourths of the children in New York City showed signs of rickets. Now, due in great . . . [to] vitamin D fortification of milk, infantile rickets in the United States is extremely rare" (Berg 1973, 109–110; see also Etheridge 1993, 918–924; and MMWR 1999h). Berg, as with most specialists, recognizes other contributions to these improvements.

At the beginning of the twentieth century, nutritional sciences were in "their infancy." The understanding that "minerals and vitamins were necessary to prevent diseases caused by dietary deficiencies" was unknown. Nutritional deficiency diseases such as rickets, scurvy, beriberi, and pellagra were thought to be infectious diseases (MMWR 1999h). By 1900, biochemists and physiologists had identified protein, fat, and carbohydrates as the basic nutrients in food (MMWR 1999h).

It was not until 1916 that "new data had led to the discovery that food contained vitamins, and the lack of 'vital amines' could cause disease." These early twentieth-century advances in science were followed by "public health policies, such as food fortification programs," resulting in "substantial reductions in nutritional deficiency diseases during the first half of the century" (MMWR 1999h). Our food is now safer, cleaner, and more abundantly available and we have greater knowledge of our nutritional needs.

Irradiation

It is interesting to note that vitamin D was first made in the 1920s the same way that the sun and our bodies do it, by radiating fats (cholesterol) with ultraviolet light. It was proudly advertised in national magazines as "Sunshine Vitamin D by Irradiation" in advertisements that touted irradiated products, including "Irradiated Evaporated Milk" (Apple 1989, 374; and Apple 1996, 33–53, 206).

With the current climate of fear, were this technique developed today, it would have not been approved for use. These phobias in large measure account for the spirited opposition to irradiation of food, which destroys micro-organisms and thereby makes food safer. Surveys show that "education plus food samples" increases public support for irradiated foods (Bruhn 1997; and Osterholm and Potter 1997). The

American Medical Association, World Health Organization, Institute of Food Processors, American Dietetic Association, and a number of other professional organizations have endorsed food irradiation. It is in use in over forty countries, and is used to protect the food eaten by astronauts. In early 1999, the Food and Drug Administration approved the use of irradiation for red meat, followed later in the year by approval by the U.S. Department of Agriculture, even though there was still opposition to it. The approval of irradiation for meat went into effect in February 2000. The FDA also approved a new "electronic beam" pasteurization process that holds promise for making our food safer. Dr. David Kessler, former head of the Food and Drug Administration, along with other scientists, leading health and consumer advocates, has strongly argued that fears of food irradiation are unfounded and the potential benefits should not be denied to consumers (Lu 1999; andSteele 1999). Anderson (1998) says, "Not to do something . . . is to condemn us as a country to suffer the consequences of . . . food borne illnesses again and again. . . . Food irradiation is a food safety tool that we as consumers should not ignore."

Irradiation could reduce spoilage or the need for fungicides for fruits, and increase the safety of meats. Lutter (1999) says, "Irradiation is extremely effective at reducing pathogens" and is the only way to control *E. coli* O157:H7 in raw meat and poultry (Lutter 1999). It "destroys *Staphylococcus aureus* and *Campylobacter jejuni,* which are together responsible for more than 2.6 million food-borne illnesses per year, as effectively as it reduces *E. coli* O157:H7" (Lutter 1999, 2275).

Irradiation is effective controlling other food contaminants as well. Lutter points out: "Reductions in numbers of viable organisms would be dramatic for other important pathogens: irradiation of meat reduces salmonella levels by factors of 10 billion to 100 trillion. It is also effective for seafood, eggs, precooked meats, and produce" (Lutter 1999, 2275).

Jane Brody (1994) presents an extensive list of potential benefits of food irradiation, comparing the opposition to a turn-of-the-twentieth-century (nineteen to twentieth century) innovation in food processing (Brody 1994): "The innovation under attack almost a century ago was the pasteurization of milk, a health- and life-saving process that myth-mongering opponents kept from commercial use for 50 years"(Brody 1994).

An organic grower is quoted as saying: "Pasteurization has only been around a hundred years or so; what do you think people did before that?" Dennis Avery (1998a, 19) provides a truly elegant response: "The answer is simple. They died young."

Notes

1. See also Milton (1993), 90.
2. See also Cohen (1987), 266.
3. Quoted in Bazzaz (1987), 248. (Avery 1998b 19).
4. There are now reports that a widely used organic pesticide, rotenone, has been found to cause conditions in animals comparable to the pathways in Parkinson's disease found in human beings (Giasson et al. 2000; Betarbet et al. 2000; and Blakeslee 2000). It is now even more imperative that *Consumer Reports* test for organic pesticides in food. Though I have reservations about the use of high dose animal toxicity studies as a final arbitrator for policy, it is interesting and informative that the electronic media is quick to publicize every problem with modern pesticides and the "consumer" groups who urge the banning of any manufactured chemical that is toxic in animals, have maintained a deafening silence on the latest findings on rotenone (Morris and Powell 2001).

CHAPTER 5

Genetic Modification and Technophobia

I t is ironic that in opposing genetic engineering of plants, the advocates of organic agriculture are inhibiting the process of creating more disease-resistant varieties of crops that would facilitate using fewer pesticides. In the United States, where public opposition to genetically modified crops is largely limited to environmental groups, their increased planting has led to a decrease in the use of pesticides (Gaskell et al. 1999; ERS 2000). Disease-resistant varieties for nonprimary crops are also essential for developing countries. Some of these crops, "native to the subtropics or tropics," have an "untapped potential for producing food, fiber, fuel, and medicine" (Moffat 1999b, 370).

An argument has also arisen that the new genetically modified crops will displace traditional varieties, adversely affecting biodiversity and making agriculture more susceptible to a form of disease plague. With or without biotechnology, farmers throughout the world will continue to turn to higher yielding, more disease-resistant varieties of crops, which means that through time, "antique" or "traditional" varieties will no longer be cultivated and in that sense, biodiversity will be further "eroded."

England had a "media feeding frenzy" over genetically modified (GM) foods, with demands for a moratorium on their use until their safety could be guaranteed. Much of the original fear of the genetically modified foods was based on an unpublished study. Nineteen of Britain's most eminent scientists, all Fellows of the Royal Society, were critical of the study and the media uproar and "called for the use of peer review rather than public opinion to judge scientific results." They added that "it is a dangerous mistake . . . to assume that all statements claiming to be scientific can be taken at face value" (ScienceScope 1999). Scientific arguments seem to have had little effect in calming the opposition.

Prior to the hysteria, there had been "attempts at objective analysis by the Royal Society, the Nuffield Council on Bioethics, the House of Lords and House of Commons Select Committees" (Taverne 1999, 31). All were "almost totally ignored by the media" and therefore had little ability to bring some sense of balance to the public discourse (31). Two studies conducted by the National Academy of Sciences confirmed what supporters of genetically modified food have repeatedly maintained, namely that transgenic crops are "merely an extension, or refinement, of the kinds of genetic manipulation that have been performed for decades or even centuries" (Miller 2000b, 1343).

One NAS study noted what many have argued, that in the history of agriculture and plant breeding, there has never been a process where the outcome was more predictable and therefore safer. "With classical techniques of gene transfer . . . we cannot always predict the phenotypic expression that will result. With organisms modified by molecular methods, we are in a better, if not perfect, position to predict the phenotypic expression" (NAS 1989, 13).[1] Dixon (1999) contrasts the "row over genetically modified foods" with the "widespread acceptance and use of many recombinant products in health care," which include human insulin and growth hormone, erythropoietin, hepatitis B vaccine, tissue plasminogen activator, several interferons, factor VIII, and antihemophilic factor (Dixon 1999, 547).

"In the United Kingdom, many people welcome medical applications of gene technology as 'good genetics' but see genetically modified foods as 'bad genetics'" (Dixon 1999, 547). In addition, transgenic animals have become vital to research into understanding disease mechanisms and devising new cures. Over "25 percent of the top 20 drugs, for example insulin, growth hormone, several hepatitis B vaccines, and monoclonal antibodies to treat cancer" are produced using genetically modified organisms (Wilson, Hillman, and Robinson 1999). These percentages will likely continue to increase as more genetically modified vaccines and other pharmaceuticals are being devised and tested. In 1999, there were twenty-two approvals for new genetically modified drugs in the United States and there were more approvals in the past five years than in the previous thirteen years combined (Coffman 2000). Coffman (2000) says,

Nearly 100 biotech medicines are . . . helping hundreds of millions of people worldwide. Another 350 drugs are in late-stage clinical trials. . . . The record 22 drug approvals by the FDA in 1999 include breakthrough treatments for ovarian cancer, rheumatoid arthritis, hemophilia, influenza and hepatitis C.

Biotechnology Opportunity for Africa

The biosafety protocol agreed upon in Montreal specifically excluded pharmaceuticals (Helmuth 2000). Florence Wambugu (1999) asks "why there should be different standards for crops and pharmaceuticals, particularly in Africa where the need for food is crucial for survival." Having "missed the green revolution," Africa cannot afford to lose the opportunity to realize the benefits of biotechnology (15). What is critical for Africa and other developing areas is that they be "stakeholders" in biotechnology and other emerging technologies so they have some control over their development and use. Many trade regulations now in place in developed countries because of pressure from environmental groups and nongovernmental organizations (NGOs), end up restricting the export of products (a new vaccine) that are wanted by developing and other recipient countries (Chambers 1999, 39–40). Ignored in the debate on genetically modified foods are those most in need of increased food production, the poor in developing countries and the "marginalized" hungry in developed countries (Serageldin 1999a, 134).

Genetically Modified Foods

Genetically modified foods that use "plant viruses engineered to produce some of the protein fragments from disease-causing organisms" and that thereby "stimulate the immune system when eaten" would no doubt be of enormous benefit to the poorer peoples of the world. A genetically engineered potato with a vaccine that protects against the Norfold virus, the most common food virus in the United States, has been successfully tested (Tacket et al. 2000). Unfortunately, many firms developing these vaccine-producing foods seem to be among the first casualties of the "Frankenfood" hysteria (Feldbaum 1999). If the leading firm in genetically modifying crops to produce vaccines abandons this enterprise, the main victims will be the poor and vulnerable peoples of the world.

Continuing Advances in Biotechnology

While the controversy raged in the popular media, the world's leading scientific journals carried peer-reviewed articles on the continuing advances in biotechnology. For example, there were articles on the way in which essential nutrients, such as iron, can be introduced into our crops by genetic engineering, the identification and isolation of a dwarfing gene that could bring yield improvements in a number of crops comparable to those in wheat and rice in the green revolution, and genetically modified potatoes that resist a wide array of bacteria and fungi including *Phytophthora infestans,* the blight that caused the Irish famine (Robinson et al. 1999; Peng et al. 1999; Abelson and Hines 1999; JIC 1999). In developing countries, some genetically modified varieties "have improved yields and stability—often by better tolerance of fungi, viruses or soil poisons—for example, for rice in China, potatoes in Peru and sweet potatoes in Kenya" (Lipton 1999). In Hawaii, genetically modified papaya is now protected from a deadly virus, and this will likely benefit developing countries (Wilson, Hillman, and Robinson 1999, 69).

None of these real and potential benefits deterred the protestors who "ripped up" experimental fields of genetically modified rapeseed and sugar beets in England, preventing the necessary scientific research on the crops that the critics are demanding as a precondition of their introduction and utilization. Though the vandals, who have been going into the fields and greenhouses, destroying food crops and trees, presume to have superior knowledge and higher moral authority than others about the dangers of genetic modification of plants, they somehow are not always able to make the proper identification and end up destroying the wrong crops or trees (Taverne 1999, 31).

Genetic Modification of Plants to Meet Nutritional Needs

Genetic modification of plants offers great promise for improving the human diet in poor countries by adding vital micronutrients to stable food plants in a process that one author calls "nutritional genomics." DellaPenna (1999) argues that "modifying the nutritional composition of plant foods is an urgent worldwide health issue as basic nutritional needs for much of the world's population are still unmet." He adds that the diet of "large numbers of people in developing countries" consists primarily of a "few staples" such as cassava, wheat, rice, and corn (maize), being "poor sources of some macronutrients and many essential micronutrients." DellaPenna estimates that the diet of "over 800 million people" lacks "sufficient macronutrients," and the dietary deficiency of micronutrients is "even more prevalent" (375).

Grains are the major staple of world food production, for feeding animals, for production of meat and dairy products, and for use in nutrition-intervention programs and famine relief. Research using "genomics-based strategies" to produce "grains with enhanced value" and in "moving trait genes into high-yielding germplasm for commercialization" also has the potential to improve the diet of humans around the world (Mazur, Krebbers, and Tingey 1999; Trewavas 1999). Many of the world's poorest people are farm families subsisting on homegrown rice and sometimes little else. Half of the world's population or about three billion people eat rice on a daily basis, a number that is expected to rise to four billion in 2020 (Bloom 2000).

Food fortification and enrichment can be almost unbelievably cheap if one has an available carrier, such as imported wheat, to deliver it. Conversely, it is often highly difficult to create a delivery mechanism to fortify and enrich subsistence to farm families, vast numbers of which are in rice cultivation where iron and vitamin A (VAD) deficiency are most prevalent. TRF (1999a) notes, "Rice plants do produce carotenoid compounds that are converted to vitamin A, but only in the green parts of the plant and not in the component of rice grain consumed by humans." As always, it is the children who are most vulnerable: "The millions of children who are weaned on rice gruels are particularly prone to VAD since they consume little else. And children in rural areas are seldom reached by vitamin A supplementation programs."

Scientists in Switzerland claim to have developed a strain of rice genetically engineered to provide enhanced iron and vitamin A (Potrykus et al. 1999; TRF 1999a,b; Ye et al. 2000). There is a vitamin A-enhanced genetically engineered rapeseed plant for the production of rapeseed or canola oil (C&I 1999). While what appeared to be bad news about genetic modification received massive media attention, the very good news of the new variety of rice received no notice (Leisinger 1999b). Instead of being a candidate for a Nobel Peace prize, the creator of the new variety was immediately vilified, "bio-McCarthyism," as have others who have made advances that help to feed humanity (Leisinger 2000, 114).

WHO considers over 230 million children as being at risk of clinical or subclinical vitamin A deficiency (TRF 1999a). Vitamin A deficiency can lead to blindness—an estimated 230 to 250 million children are at risk with 500,000 children going blind every year from it—and lowers the body's resistance to disease, increasing the mortality from diseases as diverse as measles and diarrhea (Abelson and Hines 1999, 368). Vitamin A deficiency makes children more vulnerable to infection and worsens the course of many infections. Estimates of the number of children who die each year as a result of vitamin A deficiency run as high as two million deaths (James and Krattiger 1999). Supplementation with vitamin A is estimated by UNICEF to lower a child's risk of dying by about 23 percent. VAD is also the "single most important cause of blindness among children in developing countries" (TRF 1999a). UNICEF estimates that iron-deficiency anemia is partially responsible for about 20 percent of maternal deaths (TRF 1999a). TRF notes, "The major micronutrient deficiencies worldwide concern iron . . . 1.4 billion women suffering from iron deficiency anemia, and vitamin A-deficiency, affecting approximately 400 million children."

It is difficult to imagine any other development taking place at this time that has a greater potential to enhance the well-being of women and children. Nothing in scientific advancement or in the rest of life is guaranteed, but if using genetic modification to enhance the iron and vitamin A in rice work out in practice, then it is truly exciting. The intent is not to exploit the technology commercially but to develop "the transgenic material into rice-breeding lines" and allow the International Rice Research Institute (IRRI) to distribute these "free to local rice breeders . . . and other national agricultural research centers" who will use them consistent with their environment, other laws, and customs (NB 1999). The potential benefit of vitamin A and iron-enriched rice to the world's most needy people has not been sufficient to deter Greenpeace from attempting to disrupt the effort of IRRI to make it available to the farmers of Asia. Future historians may look back upon the discovery of the dwarfing gene and the genetic modification of rice as the two great discoveries in 1999, if not of the last decade or more, that helped to feed the population of the twenty-first century.

We have entered the twenty-first century with the creation of a new nontoxic variety of grasspea that is "expected to relieve suffering from drought-related paralysis" (ICARDA 2000). An extremely hardy plant, the grasspea is usually the "last plant left standing in times of drought" and can grow in drought-prone areas that receive as little as eight inches (200 millimeters) of rain a year (ICARDA 2000). If it is taken for about three months, it can cause retardation and death in children, or a "neurological disorder" that often results in "irreversible paralysis of the legs," a condition that currently afflicts about 100,000 people in the developing countries (ICARDA 2000). Though not a transgenic, the new crop was developed using a "technique known as somoclonal variation to force the plant to mutate and to express genes that were formerly dormant" (ICARDA 2000). It qualifies as "mutant grub."

In the 1960s and 1970s, a quality protein maize was developed that had much more protein but a lower output (among other problems), such that farmers did not find it beneficial to grow (NAS 1988). Since then, hybrid corn (maize) has become the largest food crop in Africa (Byerlee and Eicher 1997). More recently, CIMMYT (Centro Internacional de Mejoramiento de Maiz y Trigo) has come up with new varieties of maize with more protein, higher yields, and greater resistance to stress such as drought. This makes it likely that hybrid maize cultivation will expand even further. It is not clear at this time to what extent these new improved varieties are the product of more "traditional" breeding or of modern biotechnology. What is clear is that farmers in Africa have found it worthwhile going to the market to buy new seed rather than planting seeds from the previous harvest and will be doing so in increasing numbers in the future, as has been the case in Asia and elsewhere. Thus, the argument against transgenic crops, that farmers will have to go into the market to buy seeds as they traditionally have not done, simply ignores the extent to which farmers were doing this with higher yielding crops before the advent of transgenics, and will continue to do so. Developed-country activists are defending farmers, without consulting them, against changes that the farmers have been making on their own. As many have noted, no one is opting to continue using the "traditional" seed from the previous harvest if the farmer finds that to be advantageous.

Currently, very promising work in bioengineering salt-tolerant plants offers another possibility for increasing world food production and land conservation by bringing high-saline soils into agricultural production, reducing the need to cultivate other marginal lands, or utilize land that now supports wildlife, trees, and other vegetation (Apse et al. 1999; Frommer et al. 1999; Wood 1998, 378; Swaminathan 2000a). Also promising is research in rice and maize plants that can grow in highly alkaline soils and can achieve higher yields using less fertilizer. Equally important is the research with "newfound genes and enzymes" that could "enable crops to flourish on metal-rich soils and help other plants clean up heavy metal contamination" (Moffat 1999a, 369). A whole range of developments in plant and animal bioengineering is improving their efficiency in nutrient intake or their ability to withstand stress such as drought. Many critical nutrients for plants, such as phosphorus, are in the soil but in a form not accessible to the plants. Bioengineering is improving the ability for plants to harvest these nutrients (DeWitt 2000b). A new rice variety altered with soya genes appears to be more parasite-resistant than traditional varieties. Another "metabolically modified variety promises significantly higher yields" (Brough 1999; Ku 2000).

Florence Wambugu, a distinguished African scientist and director of the Kenya-based International Service for the Acquisition of Agri-Biotech Applications, spoke for developing countries in attacking "opposition to gene technology as a northern luxury" (Butler 1999, 360; Wambugu 1999). She says, "The biggest risk in Africa is doing nothing. . . . I appreciate ethical concerns, but anything that doesn't help feed our children is unethical." Stated differently, "The biggest risk of modern biotechnology for developing countries is that the technology will bypass poor farmers and poor consumers because of a lack of enlightened adaptation" (Pinstrup-Andersen and Cohen 2000, 165).

Pinstrup-Andersen and Cohen (2000) see a critical need for research in biotechnology "to focus on the problems of small farmers and poor consumers in developing countries." Strong public-sector involvement is needed to prevent the development of a "scientific apartheid" in which "cutting edge science becomes oriented exclusively toward industrial countries and large-scale farming" (165). The kind of support essential to be able to mount a public/private/foundation coordinated research effort in biotechnology oriented primarily to help the poor will simply not happen as long as there is an hysterical antibiotechnology campaign to alarm the public, whose support is so vitally necessary if poverty and hunger are to be reduced and the world's population adequately fed.

Though critics of bioengineered food crops attempt to claim the moral high ground, others view it as tremendously selfish of those in developed countries who "because of misinformation" oppose what is "probably the safest technology that humans have ever invented" (MacIlwain 1999a). Another scientist, Klaus Leisinger, believes that "we have an ethical commitment not to lose time" in turning this technology to the needs of developing countries (MacIlwain 1999a). To those who argue that genetic modification of plants and animals is "playing God," Prakash (1999a) argues that those obstructing it are the ones "who are really playing God, not with genes but with lives of poor and hungry people." Sahai (1997) questions whether it is "more unethical to interfere in God's work than to allow hunger deaths when these can be prevented." Prakash points to the Vatican director on bioethics as not fearing scientific advance and then quotes from a statement by leaders of the Church of England: "Human discovery and invention can be thought of as resulting from the exercise of God-given powers of mind and reason; in this respect, genetic engineering does not seem very different from other forms of scientific advance" (Prakash 1999a,b; C of E 1999; PAFL 1999).

Derek Byerlee argues that saying that there is enough food in the world "misses the point." In most poor countries, jobs and income depend on farming, and "growth in the agricultural sector" is necessary in order to raise incomes. "It is the only way forward" (MacIlwain 1999b). Bioengineering of food crops is a vital part of that way forward. Asia, China, India, and Japan are going forward with research and testing of GM foods, each having just developed improved varieties of rice. Unfortunately, African countries do not have the per capita income of Japan or the size of China and India as a basis for providing sufficient resources to be as effective in their biotech efforts as the Asian countries. African countries will have to engage in a number of cooperative actions to take advantage of the potential of biotechnology (Weber 2000). Despite the variation in capability, developing countries must move forward with diversified biotechnology programs. The effort to create the International Centre for Genetic Engineering and Biotechnology for biotechnology research for developing countries has run into financial difficulties (Jayarsman 2000a).

According to Gordon Conway: "There are about 1,000 Third World biotechnologists working on crop varieties, mostly rice. The debate is going on in the north and we're not hearing the voices coming from the developing world" (Wrong 2000a).

Florence Wambugu (1999) offers an array of different ways in which various kinds of "agricultural biotech" are already helping African farmers and the enormous potential it holds for future development. African "farmers are benefiting from tissue culture technologies for banana, sugar cane, pyrethrum, cassava and other crops" with a variety of other "transgenic technologies in the pipeline," particularly for crop protection (15). Not to use biotechnology does not mean avoiding "exploitation," in her judgment. On the contrary, African countries must "participate as stakeholders in the transgenic biotechnology business" (16; Juma 2000; Pinstrup-Andersen 1999b).

Both critics and many enthusiastic advocates of genetically modified foodstuffs argue that most of the advances in this technology, particularly those made by private sector firms, primarily benefit farmers in developed countries (Pinstrup-Andersen 1999a, 3). True though this may be, the irrational criticism of the technology diminishes support for public funding of agricultural biotechnology research that could bring its benefits to African and other Third World farmers, as was brought to many Asian and other farmers by the green revolution. One of the potential great virtues of biotechnology is that it is "packaged technology in a seed" that allows its benefits to be realized without "changing local cultural practices" (Wambugu 1999, 16; 2000).

African farmers have long been planting hybrid maize, as maize has become the number-one food crop in Africa while higher yielding hybrid rice (15 to 20 percent higher) has been spreading throughout Asia, including about half the rice acreage in China (Byerlee and Eicher 1997; Normile 2000). They are used to going into the marketplace to buy new seeds; doing so for GM seeds would not be new and would be welcomed if the increase in usable output warranted it. Hybrids are the product of conventional breeding, which will remain an important tool in developing new food crops. I have argued that technologies "coexist," meaning in this case, the numerous techniques of biotechnology will add to our ability to develop the food crops to feed a growing world population without necessarily crowding out previous techniques that remain effective (DeGregori 1985, 39). Humans cannot intelligently overlook any feasible research and development methods that will help to provide the world's population with more food and better nutrition.

Critics use the term, "Frankenstein foods" or "Frankenfoods" and "mutant grub" to describe the food products of genetic engineering. Other fears are that GM crops would outcross with wild relatives and create "superweeds." Natural Reality threatened to release a superweed it had created (Hodgson 2000b). Widely circulated in the media have been reports that the *Bacillus thuringiensis* (Bt) modified plants threatened the continued existence of monarch butterflies but the solid, scientific criticism of that thesis was basically ignored by the media (Losey et al. 1999). Shelton and Roush (1999) found it "surprising . . . that a previous and more relevant and realistic field study" was "largely overlooked by the media" (832; CNS 1999a). Also not widely publicized by the media was the fact the study was admittedly "preliminary rather than definite" and that "it would be inappropriate to draw any conclusions about the risk to Monarch populations in the field based solely on these initial results," as noted by scientists who praised the study, finding problems without faulting its merits (Beringer 1999; Pimentel and Raven 2000).

The results of Losey's monarch study were based on the laboratory feeding of monarch butterfly larvae with leaves from milkweed plants (the larvae's food of choice) that had been dusted with the pollen from a Bt GM maize (corn) without there being a field study. The actual field studies—"ongoing monitoring"—of Bt crops found that the pollen diminished rapidly "only 3 meters from the corn field's edge" (Palevitz 1999). Field studies showed that the monarch butterflies generally did not lay their eggs on milkweed plants near cornfields. Corn pollen is relatively heavy so that about 90 percent of it lands within about fifteen feet of the corn and that about 95 percent of corn pollination has been completed before the monarch eggs begin hatching. The 1999 Bt corn crop was about 30 percent of all U.S. corn acreage and did not seem to have harmed the monarch butterfly, as it was a very good year for them, measured by the numbers arriving in Mexican sanctuaries.

Other field studies have shown Bt corn not to be harmful to butterfly species related to the monarch (Wraight et al. 2000). The maize (corn) used for control in the Losey et al. (1999) study was from an "unrelated, untransformed hybrid." Therefore, it cannot necessarily be inferred that it was the Bt toxins that were responsible for the subsequent reduction in monarch butterfly larvae maturing into butterflies (Beringer 1999a; Pimentel and Raven 2000). The "Bt toxin in pollen is below 0.1 parts per million, a level that stunts corn borers (*Ostrinia nubilalis*) but doesn't kill them" (Palevitz 1999). "Bt crops involve a couple of orders of magnitudes less exposure to toxins compared to wholesale spraying with conventional insecticides or Bt preparations" (Palevitz 1999).[2]

The organic agricultural critics of genetically modified crops might be hoist on their own petard on this one. For if Bt toxin in genetically modified maize (corn) is harmful to monarch butterflies, then the much greater magnitude of Bt toxin in the fields of organic crops might be a "couple of orders of magnitude" more harmful to them. To be consistent, the critics of GM crops should apply the precautionary principle to organic agriculture using Bt spray. They should also apply the same precautionary principle when introducing alien live species into an environment as biological pest control agents in organic agriculture (Simberloff and Stilling 1996a,b).

The biological control industry has been allowed to grow into "the most powerful under-regulated or even unregulated environmental technology in the United States," despite a growing body of evidence showing harm to nontarget species and the environment. Its proponents seek to invoke the precautionary principle to competing technologies (Lockwood 2000, 16; Miller and Applet 1993).

While some environmental groups are promoting biological control, other environmental groups are seeking a ban on the introduction of "alien species." Most agriculture after its very earliest origins involved the diffusion of domesticated plants into new environments making them alien species by definition. In our global environment, the accidental introduction of alien species that can disrupt the environment is an ever-present danger. In addition to agriculture and biological control, some alien species are beneficial, so we have to evaluate each on its own merits and not a priori (Bailey 2000). Increasingly, protective measures are being taken by countries and by international action, but disruptive species still find their way to new environments.

GM corn "has a distinct health benefit of discouraging the buildup of mycotoxins in corn, potentially dangerous human and animal toxins produced by fungi that cause plant disease" (APSnet 1999). Insects that damage plants also make them more receptive to disease invasion and serve as carriers for these disease pathogens. "Insect larvae chew on stalks and kernels, creating wounds where fungal spores can enter the plant. Once established, these fungi often produce mycotoxins" (APSnet 1999). Some mycotoxins such as fumonisins "can be fatal to horses and pigs, and are probable human carcinogens" (Munkvold and Hellmich 1999; USDA 2000). The fumonisins are associated with Fusarium ear rot, the most common ear rot disease in the Corn Belt; it can be found in nearly every cornfield at harvest (Munkvold and Hellmich 1999). The Bt corn, in resisting insect damage from corn borers, also protects against disease invasion with fumonisin and mycotoxin levels thirty- to fortyfold lower in some tests (IPMnet 2000). The associations between:

> These insects and corn diseases result from several types of host-insect-pathogen interactions . . . European corn borer larvae carry spores of Fusarium species from the plant surface to the surfaces of damaged kernels or to

the interior of stalks, where infections are initiated (Munkvold and Hellmich 1999).

Kernel rot caused by *Aspergillus* also is associated with insect damage to ears. *Aspergillus flavus* and *A. parasiticus* produce the most notorious mycotoxins in corn, the aflatoxins, and "can be passed into milk" when the infected grain is eaten by the cows, making "the economic impact of aflatoxins" even greater "than that of other mycotoxins in corn" (Munkvold and Hellmich 1999).

Live Bt is defined to be "natural" and is used in "organic" agriculture. One fear of the "organic" agricultural practitioners and environmentalists is that the more widespread use of the protein from it as a built-in pesticide for plants will hasten the development of insect resistance to Bt. In many respects, using the plant to produce the Bt toxin may be safer from both an environmental and human health perspective than using *Bacillus thuringiensis*. *Bacillus thuringiensis, Bacillus cereus, Bacillus mycoides,* and *Bacillus anthracis* are members of the same genus, and many scientists consider them actually the same species or very closely related as members of the *Bacillus cereus* group. The plasmids of Bt "seem innocuous" but are "the main part of Bt's genetic material codes for toxins that can cause diarrhoea, vomiting, muscle and kidney damage and liver failure." Because Bt has a "novel gene-swapping system that enables Bt to exchange an unusually wide variety of DNA with other *Bacillus* cells, there is the potential for spawning very dangerous strains and unleashing them into the environment" (MacKenzie 1999, 22). The genes for critical toxins are present in Bt, and we currently do not know enough about the regulators that cause these pathogens to be expressed (Agaisse et al. 1999; Bouchie 2000). Additionally, Bt has been shown to cause fatal lung infections in mice, a factor that would have immediately condemned it if it were not "organic" (Trewavas 2000). When it comes to that which is deemed to be organic, the vivid and creative imagination to potential dangers ceases to function, and the precautionary principle is simply ignored by those who would save us from perils of modern technology and science. As one cogent defender of biotechnology demonstrates, we face risks continuously. "The average human comes into daily contact with a million species of bacteria and about 5,000 viruses" (Jenkins 1999). Any of them could mutate into a deadly disease but fortunately they don't, or very rarely do, and most of us don't worry about them as we know that the real risk is not all that great.

The same groups that oppose the patenting of life forms now somehow claim exclusive use—an ownership property right—to what they believe to be a product of nature. What these groups fail to mention (or are possibly even unaware of) is that the strains of Bt toxin now in use in agriculture are themselves products of genetic transformation. "Up until the advent of genetic engineering, Bt had been a fairly useless product . . . with the exception of Southeast Asia where it was heavily used." Chambers adds that Bt "breaks down in sunlight in its native form" (Chambers 1999, 39). For a history with a superb bibliography of the identification and discovery of various strains of Bt, the evolution of its usage, and the genetic manipulation of Bt for use in agriculture, see Felsot (2000a,b,c). For a study of the safety of using Bt, see WHO (1999e). Says Chambers (1999, 39), "Only a few Bts . . . were of application in agriculture . . . we did not have the expertise of biotechnology to explore the variety of strains . . . and develop them into a number of different, highly useful genes and products."

International Trade

Genetic modification is increasingly becoming an issue in international trade, as countries and/or trading blocs try unilaterally to prevent the importation of foodstuffs from genetically modified plants (or beef from hormone-fed cows) by treaty, in violation of existing trade agreements that prohibit such barriers to trade unless they are based on scientific evidence. Developing countries without the technical expertise on the issue are seeking guidance from organizations such as CGIAR (Consultative Group on International Agricultural Research) (MacIlwain 1999a). The Rockefeller Foundation, which funded the research on nutritionally enhanced varieties of rice, is worried that public opposition to bioengineered foods in developed countries might make it more difficult to introduce improved varieties of bioengineered crops in developing countries where they are most needed (Lehrman 1999; Normile 1999; Conway 2000).

Few could object to some form of labeling or a "prior informed consent" arrangement (emphasis on "informed," not sharing phobias), provided it is not simply a ploy to mire the process in bureaucracy and significantly raise the cost of the food. One of the many purposes of genetic modification is to increase production and lower the cost, making food more available to the needy. Certainly, there would be no objection if private groups wish to band together to establish standards and organize their own labeling. This is done all the time by religious groups who set standards and arrange for labeling as kosher or nonkosher or Halal or non-Halal. Government is and should not be involved in the process except in case of fraudulent labeling. The religious groups neither seek, nor would they welcome, government involvement in a process that has little if any meaning to the rest of the community.

The consensus is and has been that labeling is for objective, scientific information, of use to all consumers. In the context of the current debate over genetically modified foodstuffs, labeling would imply a scientifically verifiable difference in terms of health and nutrition when there is none. It would serve to validate the arguments of the opponents of genetic modification in the minds of the public, even though the opponents have failed to do so in the scientific journals. As a practical matter, we cannot accommodate on food labels the personal preferences of every religious, ethnic, or other group that has dietary prohibitions. Basic information is provided to everyone; the rest is up to each group. To require labels to state "GM" or "GM free" makes no more sense than to require all labels to state that they are kosher or nonkosher.

A moratorium would not be a time period in which further inquiry and testing would take place, since there are no scientific results that could even remotely guarantee safety for those devotees of the precautionary principle. After Canada formed a prestigious scientific group to study the safety of GM food, a spokesperson for Greenpeace indicated that the controversy would continue regardless of the panel's findings (Greenpeace 2000).

Greenpeace has joined with other groups to file a class-action suit against Monsanto to stop its development and marketing of genetically modified agricultural plants, and it and sixty-four other plaintiffs are suing the Environmental Protection Agency (EPA, arguing that it was unlawful for the EPA to approve genetically modified crops that produce the same toxin as that produced by the bacterium *Bacillus thuringiensis* used in "organic" agriculture. It should be noted that there is no allegation that the crops or foodstuffs are unsafe but that there was inadequate testing for safety. This is a clever but not entirely honest strategy, since the evidence that could be brought in defense of the safety of the genetically modified plants would overwhelm the case against in an American judiciary that is increasingly wary of the "junk science" that the plaintiffs bring to court in an attempt to prove their case. One can expect that the case will be marketed to the public as one of food safety (with the implicit claim that GM foods are "known" to be dangerous), not of testing procedures.

Both a moratorium and restriction on trade would simply provide greater opportunity for those opposed to genetically modified foods to whip up more hysteria and have them banned completely. They argue, in the international trade debate, for full disclosure and prior approval on genetically modified foods, that it is also an issue of consumer choice. It is an ironic use of language that banning a product that others of us wish to consume for which there is no *verifiable scientific evidence of harm to humans* is not a restriction of choice nor are highly bureaucratic procedures for international trade in food, which would make it more expensive and less available to those who need it most.

There are serious scientific concerns about foodstuffs having genes that are inserted from other plants—such as fatal food allergies. Applying the label of "Frankenstein foods" immediately polarizes and trivializes discourse on the issue and makes it difficult, if not impossible, to reach public understanding and policy formulation that allow us to realize the benefits of the technology and avoid the potentially deleterious outcomes. The regulations in place for genetically modified foods make them one of the most carefully regulated consumer products (Gasson 1999).

War of Words

The controversy over genetically modified food crops is indicative of the difficulty in using the scientific method for public understanding of complex issues when it is in opposition to skilled propagandists working for strongly motivated ideological groups. The opponents of genetic modification of food crops thus far have won the war. It was won before the first battle was fought when the technophobes came up with a variety of scare terms—"terminator genes," "frankenfoods," and others. The "Gene Use Restriction Technology," which critics called the "terminator gene," was "merely a patent claim and has not yet been produced. This device to prevent the formation of fertile seed from a GM crop, would also prevent the spread of the inserted gene to other plants" (Lachmann 1999). Since critics had voiced concern that genes from genetically modified plants might cross with wild varieties, one would have thought that a functioning "terminator gene" would be seen as a desired fail-safe device. Tests on genetically modified maize have indicated that normal agricultural practices make cross-pollination with unmodified varieties of maize unlikely (Squire et al. 2000, 50). The issue of gene flows is complex. It occurs with traditional crops also; it is not likely to occur where there are "no native species with which to cross-breed," and it is not "invariably undesirable" (Hillman 2000, 39). With the antitechnology zealots, it is not always clear what they want, only what they are against.

Those convinced that GM food was dangerous simply ignored the fact that unknown millions of American and other people around the world had been eating genetically modified foods for several years without "one example of any identifiable medical condition induced" in those "who have consumed such material" (Malcom 1999). Initially there was no science, good or bad, merely fear of the unknown. Despite evidence for the safety of the foods, there was the implicit and explicit demand for what was scientifically impossible, namely, absolute proof that no harm could ever result from genetically modified foodstuffs. In the public domain of tabloids and television, a reputable scientist's cautious respect for the limits to the available knowledge is no match for the certainty and shrill rhetoric of the ideologues. For the ideologues armed with the precautionary principle, scientific inquiry is relevant if it appears to support your cause. The gods seemed to smile on the opponents of genetic engineering of food crops as a result of the research on genetically modified potatoes.

> An entirely sensible study had been undertaken to see whether lectins which make some plants unpalatable to insects could be introduced into other plants for the same purpose. The study used potatoes only to make the experimentation easier. These particular potatoes were never intended to be developed as a food crop (Lachmann 1999; Bowden 1999).

One of their scientists announced on television last autumn that feeding these transgenic potatoes to rats had caused abnormalities of organ growth and had damaged their immune systems. . . . These remarks were seized upon by the tabloid press and engendered an hysterical reaction that has not died down (Lachmann 1999). As the British tabloids rushed to judgment, deeming genetically modified foods to be unsafe, scientists were appalled that research results would be released at a press conference without first presenting the data on which the results were based and submitting them in a paper to be peer reviewed. The Rowett Research Institute in Scotland "arranged for an audit of Pusztai's results, which concluded that his data were 'too inconsistent' to support his conclusions" (Masood 1999a,b).

The Royal Society sought the data from the researcher to have them peer reviewed, following a standard scientific practice of anonymous reviewers. A "working group" of the Royal Society had the data peer reviewed by six independent, anonymous "referees with a range of expertise" and were "very careful to review only the quality of the science and not to comment on the actions of anyone involved" (Bateson 1999). They found that the "study was based on flawed design, execution and analysis" (Loder 1999a,b; RS 1999a,b).

The scientist, who showed so little regard for time-tested methods of science, in the way that he first made his results public, compounded the mischief by implicitly attacking the system of peer review itself in referring to the "unsolicited report of the Royal Society," and attacking it as being the product of "clandestine peer review." Without knowing who was on the peer-review committee, he implied that its members were biased "because many influential committees are redolent with advisers linked to biotechnology companies" (Ewen and Pusztai 1999a).

As the details of the research and the methods used were revealed, further doubts as to its validity emerged. When the results were finally published, they drew the following careful criticism:

> The experiments done by Ewen and Pusztai were incomplete, included too few animals per diet group, and lacked controls such as a standard rodent diet containing about 15 percent protein (lactalbumin) as a balanced source of amino acids and a test diet with potatoes containing an "empty vector" (Kuiper et al. 1999).

An additional shortcoming with the study was that: "the diets were protein deficient; they contained only 6 percent protein by weight. There is convincing evidence that short-term protein stress and starvation impair the growth rate, development, hepatic metabolism, and immune function of rats." Furthermore, "the results are difficult to interpret and do not allow the conclusion that the genetic modification of potatoes accounts for adverse effects in animals." The results were what one would expect from the rats' diet. Ingestion of potatoes "may be associated with several adaptive changes in the gut because of the low digestibility of raw or partly refined potato starch. In rats caecal hypertrophy is a common response to short-term feeding of various poorly digestible carbohydrates" (Kuiper et al. 1999).

There were many other difficulties with the research. The researchers claimed that it was not the potatoes that were responsible for the differential impact upon the rats but the "construct," the mechanism by which the potatoes were genetically modified. When the published data were examined it was found that "the content of starch, glucose polymers, lectin, and trypsin and chymotrypsin inhibitors in GM potatoes differed from that of the parental line" (Kuiper et al. 1999).

As leading scientists with prestigious scientific organizations and societies offered critical analysis of the research, the public, in a distrust of authority borne of earlier incidents, reacted more strongly, and opposition to genetically modified foods grew and spread across Europe. The establishment being against you proves that you are right but when the prestigious (establishment) medical journal the *Lancet* agrees to publish the research after numerous revisions, that also proves you right and was shamelessly seen by environmental groups as being a vindication of their position. They claimed vindication despite the fact that the editor of the journal, in defending his decision to publish, clearly stated: "This is absolutely not a vindication of Dr. Pusztai's claims. But we can now draw a line under the phoney debate we have had for the last year" (Horton 1999). Further, the editor stated that "Pusztai had recklessly made claims about his data a year ago" and argued that a benefit of publication was that Pusztai "has had to retract his original claim because his data, which we've published, absolutely don't show that genetically modified foods stunt the growth of rats" (Loder 1999b, 731).

Only a committed ideologue could consider this to be a vindication. Reading the editor's justification of the decision to publish, one searches in vain for a clear statement that this piece would have been published on merit alone, even if there had been no controversy surrounding it. The editor, Richard Horton, defended the peer-review process "by six specialist advisers . . . who had several requests for clarification about the design of the study, the laboratory methods used, and the statistical tests applied." We have some who advised rejection and others who encouraged publication. One reviewer argued "that the data were 'flawed'" but wished to see it published to put it in "the public domain so that fellow scientists can judge for themselves

. . . if the paper is not published, it will be claimed there is a conspiracy to suppress information" (Horton 1999).

To some of us, public hysteria was a factor in the publication of the piece. The editor speaks about the need for "accountability" and about public perceptions of risk. "Risks are not simply questions of abstract probabilities or theoretical reassurances. What matters is what people believe about these risks and why they hold those beliefs" (Horton 1999). If a group can whip up public hysteria about some possible risks, real or otherwise, then "what people believe about these risks" becomes at least a partial reason for peer-reviewed publication.

The Kuiper, Noteborn, and Peijnenburg (1999) critique in the *Lancet* of Ewen/Pusztai research quoted above is based upon the article published in the same issue of the *Lancet*. The Ewen/Pusztai article is silent on some of the more sensational, previously made claims, such as stunted organ growth or damage to the immune system. Yet, activist groups in their campaign against GM foods are still making the claim of "damage to vital organs and their immune system." The rats in the control group were adversely affected by the diet, as the critique indicates, but the sole finding was that the rats fed the GM potato diet suffered more damage. Many scientists thought that publication of the research set a dangerous precedent in that it rewarded researchers and interest groups for fomenting public hysteria. Many, too, thought that it damaged the reputation of the journal in which it was published. "To me, it tarnishes the reputation of the journal that publishes it," said Floyd E. Bloom, editor of *Science.* "If you're just going to take it because it's controversial, well, there are a whole lot of controversial things" (Pollack 1999). Marcia Angell, editor-chief of the *New England Journal of Medicine,* asked "When was the last time [the *Lancet*] published a rat study that was uninterpretable? . . . This really was dropping the bar" (Enserink 1999b; *Lancet* 1999b; Marwick 2000a). Lancet editor Richard Horton, in defending the publication of the Ewen/Pusztai article, noted these critics.

The controversy over genetic modification of foods is didactic in other ways. Many scientists argue that any time one engages in plant breeding by crossing different varieties, one could emerge with combinations of genes whose impact on human health is unknown. The outcome of GM is more limited and therefore should have fewer unknowns, rather than more, as the opponents of GM claim. New plants or existing crops grown under different circumstances might have outcomes that are dangerous to human health. Trewavas and Leaver

(1999) give two examples of many actual outcomes. One concerns light-activated psoralen in insect resistant "non-GM celery" that causes chloracne, sometimes severe. They stated, "Cool weather-induced toxic accumulations of solanine caused the withdrawal of the non-GM Magnum Bonum potato line in Sweden."

The standard varieties of celery with other umbelliferous vegetables (containing furocoumarins) such as parsnips, carrots, and dill (with other foodstuffs, cloves, and figs) cause a serious case of phytophotodermatitis or skin eruption for those who pick them when the sun is bright (Lutchman, Inyang, and Hodgkinson 1999). We have long known that potatoes were loaded with various toxins. It may be argued that contemporary genetic modification of plants is a continuation of traditional plant breeding; many proponents of GM foods do accept this.

If in fact there was a break from "tradition," it was at least a century ago with the work of Luther Burbank who in his time was accused of playing God by crossing species and creating new ones. The use of the alkaloid colchicine ($C_{22}H_{25}NO_6$) in plant breeding meant that diploids and haploids could be crossed. Swaminathan (2000b, 38) notes, "In the 1930s and 1940s, several new methods of chromosome and gene manipulation were discovered, such as the use of colchicine to achieve a doubling in chromosome number."[3]

In addition, hybrid vigor in maize and other crops was widely commercialized. This was to be followed later by the use of chemicals such as nitrogen mustard and ethyl methane sulphonate to induce mutations, techniques like tissue and embryo rescue to get viable hybrids from distantly related species (Swaminathan 2000b, 38; and Persley 2000, 4).

The identification of the "double helix structure of DNA (deoxyribonucleic acid)" by James Watson and Francis Crick in 1953 accelerated the transition from a Mendelian approach to plant breeding to one of molecular genetics (Persley 2000, 4).[4] Swaminathan rightly argues that this "capsule" history of twentieth-century plant breeding represents a "continuum with each new generation, taking our understanding of the complex web of life to a higher level" (Swaminathan 2000b, 38). Since the 1950s, plant breeders have been exposing seeds to "heavy doses of gamma rays from a nuclear source, resulting in severe, widespread and random genetic change." Most of the mutant seeds are "useless but occasionally" there are alterations that "lead to more productive strains," which historically have "gone straight into the field with no safety testing," even though it was not known "which of the genes have mutated or how" (Hodgson 2000a). As of 1995, the FAO/IAEA (Food and Agriculture Organization/International Atomic Energy Agency) Mutant Varieties Database "included 1790 cultivars involving 154 plant species . . . in more than 50 countries, of which the 'top six' were: China, India, the former USSR, the Netherlands, Japan and the USA" (Harten 1998, 17). It is fair to say that it is virtually certain that all those protesting "mutant grub" in Europe and North America have been consuming it in various forms for some time with no known ill effects.

Mutagenesis, or "gene change by deliberate exposure to intense nuclear radiation or chemical mutagens," has the least predictable outcome of all forms of plant breeding, yet has until now escaped the label of being genetically modified, since these techniques have been around for a half a century. Somehow, those who oppose GM foods are able to proclaim crops produced by the various methods noted above to be conventional with many "being used by all farmers, including organic ones" (Hodgson 2000b).

Also generally accepted as GM free are many food products such as cheese made with genetically modified microorganisms. The Organic Trade Association (OTA) itself is not clear whether some of the current production inputs (including enzymes) into organic foods may not meet the new standards ("GM free?") to be so labeled.

If in fact cheese and other products that have been sold as organic violate, for whatever reason, the new codes for being organic (as demanded by its devotees), then one wonders why go to the cost and trouble to make the changes if there is not even the slightest hint or suggestion of there having been any harm from consuming them? The OTA blames the USDA for not having provided the funds for a task that should be the responsibility of the OTA itself. If there is no evidence of harm, why should the public taxpayers pay to gratify the fetishes of a single group? The organic anti-GM food consumers seek to impose the cost of their ritual beliefs on the rest of the community by having the public pay for the enforcement of their regulations or pay the substantial cost of segregating GM and GM-free crops, or by denying the public the choice of products that the public wishes to have.

Other GM-free products include detergents that have genetically modified enzymes for removing dirt and stains, replacing more environmentally destructive ingredients. Hapgood (2000, 49) presents a diverse list of the many current and future uses of bioengineered products.

In attempting to appease the anti-GM hysterics, governments in Europe started making obligatory noises about tougher food safety regulations. Writing in the London newspaper the *Express,* Mo Mowlam, the minister for the Cabinet Office, bragged that the government "had created one of the toughest regulatory systems in the world for GM foods" (Linden 1999). Her comments were more revealing than was intended as to the extremeness of the regulatory response. She says, "If Elizabethan England had the sort of regulatory system we have now, there is no way the potato would ever have been introduced into this country" (Linden 1999).

Many of the world's most important cultigens would have never been introduced outside the original area of domestication, had the current hysteria been operative and had similar restrictions (in terms of the knowledge of the time) in the name "food safety" been in place. The world's citizens would be unimaginably poorer both in our ability to feed ourselves and in the choice of foods. The poor in developing countries and the world in general will be far poorer if the current hysteria is allowed to prevail.

Some GM crops did in fact have the deleterious potential claimed by their opponents, including crops with wild progenitors that could have been transformed by the new GM crop, and food crops that had the allergenic properties of the transferred gene. Existing procedures identified these difficulties in the testing phase, and their development was correctly terminated. Since over 90 percent of food allergies result from specific proteins in eight foods: peanuts, tree nuts, milk, eggs, soybeans, shellfish, fish, and wheat, procedures are in place to require tests for allergenic responses in foodstuffs altered with genes from these sources.

Rather than having been insufficiently tested, as the critics claim, genetically modified crops have been subjected to far more testing than would be warranted by any consideration of potential danger. It was the major developers of GM crops that lobbied in the United States for strict regulation that was costly to the producers, intending to make it difficult for independent entrepreneurs, small biotech firms, and university researchers to compete against them (Miller 1999, 2000a). To Miller (1999), the big firms scored a "Pyrrhic victory" in keeping down competition but thereby "fed the anti-technology mythology that has poisoned views of the consumers." However we may view the transgressions of the biotech seed industry, they "pale beside the actions of their ideological opponents." In the process, we are all losers (Miller 1999).

It should also be noted here that of all the different ways humans have bred plants during the entire history of agriculture, the genetic engineering of gene splicing has the most predictable outcome with fewer unknowns; consequently, in addition to all the other benefits, it is also the safest. The hysteria over "genetic engineering" and the application of the "precautionary principle" to it and not to its alternatives is therefore not only irrational but also incomprehensible. Farmers, seed companies, and consumers are all benefiting with the farmers thus far receiving the largest share of the gain (Falck-Zapeda, Traxler, and Nelson 1999; 2000a,b).

The words of Patrick Moore, a green activist and a founder of Greenpeace, are instructive. He accuses Greenpeace of "abandoning science and following agendas that have little to do with saving the earth" (Bond 2000, 77). Moore cogently argues that "I believe we are entering an era now where pagan beliefs and junk science are influencing public policy. GM foods and forestry are both good examples where policy is being influenced by arguments that have no basis in fact or logic" (Bond 2000, 77).

Moore recognizes the need for intelligent caution and concern, not irrational fears and hysterical opposition to potentially beneficial technological change. He adds:

> Certainly, biotechnology needs to be done very carefully. But GM crops are in the same category as oestrogen-mimicking compounds and pesticide residues. They are seen as an invisible force that will kill us all in our sleep or turn us all into mutants. It is preying on people's fear of the unknown (Bond 2000, 77).

The Precautionary Principle

The genetic modification debate has given new life to what is being called the "precautionary principle," which claims that if the possible adverse consequences are catastrophic, then preventive action should be undertaken, in spite of lacking evidence to support the prediction. This may sound like a prudent course of action, but it would in fact hold public policy hostage to those with vivid imaginations who are most vocal in proclaiming their phobias. It has been called "the categorical imperative of environmentalists," which translates to mean "thou shall not tolerate even a risk of risk" (Bate 1997). Noting a similar extreme formulation to the precautionary principle by a conservation organization, Foster, Vecchia, and Repacholi (2000, 979) argued that this interpretation meant that there could be no technological progress. Even if the technology can be used safely, Monbiot (2000) would have us oppose its introduction because scientists and other humans, being inherently corrupt, will use it in a destructive manner. It is hard to imagine new technologies or scientific discoveries being allowed under these extreme perceptions of humanity; our capacity to use our technology intelligently would be squelched. As the editor of the website junkscience.com said, "had Monbiot been around in prehistoric times, he probably would have discouraged the use of fire."

The precautionary principle is often defined as "absence of evidence is not the same as absence of risk." What this really says is that the proponents of the principle have lost the argument on the evidence (otherwise they would argue the evidence), so they argue that we should follow their policy prescriptions anyway. Stated differently, if our fears and phobias are right, we are right, but even if we are wrong, well we are still right: it's my policy, right or wrong. In a similar instance on a study that found it unlikely that genetically modified food crops would create the much feared "super bugs," a spokesman of Friends of the Earth found the study to be "interesting" but he didn't think that it was "the definitive piece of research." This came from a movement and organization that consider any study, no matter how shoddy, to be definitive if it supports the anti-GM foods phobia (Arthur 2000).

As with many antiscience and antitechnology phobias, the precautionary principle would have us ignore the continuing benefits derived from using a technology. In capturing public attention and valuable resources for trivial or nonexistent problems, it would prevent these resources from being used to address clearly identifiable problems where the benefit-to-cost ratio is much higher (Breyer 1993). The precautionary principle applied to food production "would require that we grant legitimacy to the belief that scientifically unwarranted concerns for environmental safety take absolute precedent over providing a population with the means to feed itself" (della-Cioppa and Callan 2000).

Politically, the precautionary principle is offered as a radical, even revolutionary, doctrine when in reality it is way beyond conservative in being fundamentally reactionary and elitist. More than even the most conservative doctrines, it assumes that the status quo is privileged and free of danger. This may be fine today for comfortable elites, but it adversely impacts the disadvantaged, particularly those in poorer countries who need technological change to raise their living standards and improve their lives. If one looks at some of the innovations, such as immunization, that have contributed so mightily to bettering the human condition, one has to ask how many would have been allowed to happen, had the precautionary principle been operative. The European Community's endorsement of this principle is more a capitulation to street mobs and public hysteria than to any reasonable concern for public safety (ComEurCom 2000).

As with the authors quoted above, the precautionary principle assumes that the risks are there but are yet to be discovered. In an article titled "Absence of Certainty Is Not Synonymous with Absence of Risk," Cairns (1999) states: "Unrecognized risks are still risks; uncertain risks are still risks; and denied risks are still risks." This is indisputably true, provided there are in fact undiscovered risks. The author's statement simply assumes what is yet to be proved. Assuming the outcome of scientific inquiry before the results are known is becoming all too common, when a person or group wishes to use them to advance its agenda. In a dispute over the scientific basis for European trade sanctions on beef from cattle that had been fed hormones, an official for the European Union argued in early February 1999 that a scientific study would produce findings in May of 1999 that would allow the EU to "fully implement" its obligations under the World Trade Organization agreements (Williams 1999). The question is, if one knows the results of a scientific study a priori (actually seventeen such studies were said to be in process), why bother to carry it (or them) out? "After 10 years and two WTO rulings against it, the EU continues to search for the 'right' scientific evidence to support a political prejudice against beef raised with growth hormones" (Eizenstat 1999; Winestock 2000; AgWeb 2000). The United States is not opposed to the use of the precautionary principle in trade agreements, as long as it is not used as an excuse for restraining trade where there is a lack of any substantive evidence for real harm.

The precautionary principle assumes that there are risk-free alternatives. In all human endeavors, there are no risk-free actions; even inaction has a risk factor. Further, the precautionary principle purports to be science but in fact is more on the order of theology. It implies that the realm beyond what we currently know is the exclusive preserve of the critics of technology. In other words, if the known evidence does not support your phobia, then the evidence must lie in the great unknown. The position claims that all the possible unknown dangerous actions will be from technological practices while the less technologically advanced, presumably more "natural" alternatives, carry no unseen dangers. What it means is that no matter how much we push forward the frontiers of knowledge and demonstrate that there are no proven dangers to a product or practice, there will always be the unknown dangers in the great beyond that can be called upon in an attempt to impede intelligent, problem-solving action. These are assumptions that have no basis in fact or human experience.

Frank Furedi (1997) says that by adhering to the precautionary principle, we are "institutionalizing caution. . . . By institutionalizing caution, the precautionary principle imposes a doctrine of limits. It offers security, but in exchange for lowering expectations, limiting growth and preventing experimentation and change" (9).

To paraphrase Winston Churchill, the precautionary principle will provide neither security nor growth and progressive change. To some readers of this book, further growth may seem unnecessary, but to those who have not yet received the full complement of the benefits of modern technology, further growth is a necessary condition for the expansion of life, both quantitatively in terms of life expectancy, and qualitatively in the many dimensions that I discuss throughout this book. For many environmentalist groups, the precautionary principle is primarily a political tactic and not a principle in any meaningful sense of that term. On the question of genetically modified foods, attempts to add to knowledge by field-trial studies have been disrupted by organized acts of vandalism. A meeting of four hundred delegates assembled by the OECD to explore the issue of GM foods was also condemned in advance, even though representatives of environmental advocacy organizations were invited as participants. No matter how much evidence is gathered with no evidence of harm, the only conclusion some groups will accept is one that is in line with their preconceptions.

We can measure the very real costs to children if we fail to continue to improve food supply to reduce hunger and meet the food needs of a growing population. For children, Beachy (2000) says that:

- Malnutrition in utero leads to increased diabetes, hypertension, and heart disease.
- Malnutrition in utero can cause effects two generations subsequent to the mother, with impacts on intelligence and learning.
- Low calorie intake leads to kwashiorkor, marasmus, edema, and other conditions.

- Vitamin A deficiencies can lead to blindness; Folic acid (a B vitamin) deficiencies reduce intelligence.

For GM organisms, their "methodological framework" would be based on the "stratification of organisms into risk categories according to the consensus judgments of independent scientific experts" (Miller and Conko 2000).

Notes

1. Quoted in Miller (2000b), 1343.
2. Quoting genetics professor Richard Meagher.
3. See also Persley (2000), 4.
4. For an overall survey of mutation breeding, see Micke (1999).

CHAPTER 6

Technophobia, Pollution, Pesticides, and DDT

Part of the difficulty of discussing issues of modern technology and science is that public discourse is being driven by emotional language. To some, the words *radiation, nuclear energy, pesticides,* and *chemicals* connote something evil and dangerous, while the words *natural, organic,* and *solar* connote something true, good, beautiful, clean, pure, and safe. One theology gives rise to its opposite, where nuclear becomes people's energy salvation, and solar implies all that is impractical. The problems of sufficient energy, feeding the world's population, and preserving the environment continue, but our ability to deal with them is hampered by polarization and dispute over symbols rather than over meaningful, substantive ways of solving problems. Not only is the discourse overflowing with code words, but also there are built-in mechanisms for rejecting any scientific conclusions that disturb preconceptions. They include impugning the integrity of the researchers or suggesting that the study is "flawed" or incomplete.

In a sense, all scientific inquiry is incomplete in that it does not completely close out inquiry, but we can reach a level of confidence to serve as a more than sufficient basis for public policy and private action. We have seen that many are advocating the precautionary principle. Namely, even if you can't definitely prove that an event such as global warming will occur or that it will be catastrophic if it does, the very magnitude and possible irreversibility of the outcome renders it prudent to take preventive action. Since few of these projections can be completely proved or disproved, there is some merit to the principle as long as the evidence for the adverse outcome is substantial and clearly leads over alternative explanations, provided there are courses of action that are not prohibitively expensive, do not pose other dangers, or are worthwhile in their own right. Many are advocating precautionary actions for all of their phobias, regardless of the evidence.

To some, technology and science are guilty until proven innocent beyond any possible doubt. Since it is not in the nature of scientific and technological inquiry to provide absolute and total assurance, then technology and science are doomed to be forever guilty (Russell 1990, 13–14). For example, Altman (1990), quoting environmental critics he interviewed, says that if studies find that increases in cancer are "not statistically significant" as a result of the accident at Three Mile Island nuclear power plant, they are immediately denounced as "flawed and misleading" because they did not reach the conclusions desired by the critics. Such simplistic responses ignore the enormous complexity of determining what is or is not toxic in the environment (Foster, Bernstein, and Huber 1993).[1]

Similarly, the "National Academy of Science's Committee on the Biological Effects of Ionizing Radiation and the United Nations' Scientific Committee on the Effects of Atomic Radiation examined the data and concluded that the genetic damage to a child conceived after parental exposure under most circumstances (save a nuclear war or a nuclear accident more severe than Chernobyl) is less than to a child whose mother smokes or drinks during pregnancy (Schull 1990, 271).

Phobias of the Affluent and Third World Poverty

Phobias about food and manufactured chemicals are relatively harmless in the context of affluent countries. We can afford our fetishes but the poor cannot. When NGOs and other dedicated ideological groups work through their governments in developed countries to impose their fears about foods and chemicals on poorer countries, the result is often harmful. Fortunately, food phobias in developed countries do not extend to vitamins and food fortification, which turn out to be the basis for some of the most effective life-saving interventions in economic development.

It may be true that the successes of agriculture in developed countries have made most food fortification unnecessary. However, throughout the less-developed world, iodine is still necessary for control of goiter and mental retardation (Giroud 2000). Various programs (as well as continuing study and experimentation) are ongoing in vitamin A fortification to eliminate blindness caused by vitamin A deficiency and to reduce deaths from measles, dysentery, and respiratory diseases. Few of the organic food elitists are aware of the potentially devastating impact of their legislative proposals upon the poor in developed countries and even worse upon the poor in less-developed countries (Borlaug and Dowswell 1988, 14; Smil 1991, 593). Many affluent urbanites, who wouldn't know the difference between a cow patty and a rice paddy if they stepped in one, have become "experts" on agriculture after reading tracts written by those who know virtually nothing about the subject.

From the perspective of developing countries, legislation in developed countries that attempts to limit the production and export of chemical pesticides is another instance of shifting the burden of environmental protection to those least responsible for creating the problem, who happen also to be those least able to afford the remedies. In many instances, it is questionable whether the proffered legislation is dealing with scientifically verifiable health and environmental concerns or the phobias of an affluent antitechnology elite. Whether well intentioned or motivated by crass vested interests, most of these efforts are fundamentally misguided. Even were alleged carcinogenic effects of these pesticides real, there may be different antipoverty trade-offs in developing countries from those for more-developed countries.

While in India, I was shown an improved chula (a small traditional cookstove) that would require women to gather wood only three times a week rather than each day as done previously. Open fires such as cookstoves produce a variety of particulate matter and toxic chemicals, including benzo-b-pyrene, which is considered one of the most carcinogenic substances produced by industry or by everyday activities such as burning. In most cases, increased efficiency reduces pollution. In this case, the slow burn made the cooker a more efficient pyrolizer. When I noted that they were more efficiently producing carcinogens, the reply was that most of the users wouldn't live to an age for the probable increase in cancer to occur. In other words, the increase in risk was more than offset by the benefit. This might not have been true if the chula also increased the output of eye irritants and particulate matter that raised the chances for blindness and respiratory diseases.

Any reduction in particulate matter would provide a trade-off to the carcinogenic effect. The *Economist* (1994, 15) editors note,

> According to the Tata Energy Research Institute (TERI), in New Delhi, some 6m Indians die each year from acute respiratory infections caused by smoke from their fires. With inefficient stoves, poor families often use five times as much energy to cook a meal as a European family does.

With economic development and increasing income, "people soon switch to commercial fuels, which can be burnt more efficiently" (Economist 1994, 15). Six million deaths might be a high estimate, since it would constitute close to two-thirds of those who die each year in India of all causes. Whatever the true number, whether six million or smaller, it is a very large number of deaths, with many suffering a debilitating illness from chronic obstructive pulmonary disease (Wilairatana and Looareesuwan 2000). Recent UNDP (1998, 69) figures give 2.2 million annual global deaths from burning traditional fuels, that create toxic smoke that fills the houses. A kilogram of wood provides only one-tenth of the heat that comes from burning liquid petroleum gas (Barnes, van der Plas, and Floor 1997, 3).

It makes sense for Americans to take precautions against benzo-b-pyrene and other likely carcinogens, as we have more technological options and live to an age where the cumulative probabilities of carcinogens become significant.[2] Globally, it is estimated that one billion people, mainly women and children, are exposed to intense indoor pollution (WHO 1997). "Acute respiratory infections caused by indoor air pollution from biomass burning contributes to some 4 million deaths annually among infants and children" (Hansen 1992, 121; Barnes et al. 1997, 4). The estimates of deaths from respiratory diseases provided by different agencies are not always consistent; it depends on whether the deaths are from a specific source, such as indoor pollution, or of a particular group or infants and children. In any case, the number is large, significant, and by some estimates the greatest single cause of death in developing countries.

Poverty is the worst form of pollution for those who are condemned to it. The poor (mostly women) cook over open fires, thus suffering respiratory illnesses, eye irritation, and blindness, and cancer, if they live long enough. The interior of huts where these fires is is far more polluted than the air breathed by all but a few industrial workers in developed countries and developing countries. The poor get sick and die from unclean water. Estimates are that in urban areas, air pollution is reduced progressively as income passes a threshold (about $3,000 per capita) where the use of better technologies and pollution-control devices becomes affordable. Protectionist actions that restrict imports in the name of preserving the health of workers in developing countries could end up keeping the alleged beneficiaries both poor and polluted.

For many banned chemicals, there is substantial scientific debate as to the probable dangers to human health or to the environment, or whether they are harmful at all. Even if the worst-case scenarios are accepted as scientifically valid for a pesticide, there may be legitimate reasons for using it in a developing country. In most humid tropical regions, it is difficult, if not close to impossible, to grow many vegetables without insecticides and fungicides. Before modern chemical pesticides, farmers throughout the world used a variety of toxic substances such as arsenic. As we have shown, increased vegetable production and *consumption* have prevented blindness and saved lives. Further, fungal infestation of vegetables and grains, even when it is not visible, can be far more toxic and carcinogenic than all the fungicides, pesticides, and other chemical products of modern industry. In laboratory tests, for aflatoxin B secreted by *Aspergillus flavus,* "the dose-response relation for aflatoxin-induced carcinogenity is . . . one-ten millionth that required by saccharin . . . needed to increase the incidence of tumors in experimental animals" (Rodricks 1992, 163).

Aflatoxins have been a cause of misery and death throughout our history and remain a strong cause of disease and death in developing countries. We in developed countries are routinely protected against the worst fungal infestations in our foods, such as groundnuts (peanuts) or grains such as maize (corn); people in poor countries are not. If invasion by microorganism occurs, the plant will manufacture and release toxins to protect against the invaders. Some of these toxins are arguably worse for our consumption than the chemicals used to protect the plant, many of which have been shown not to be deleterious to human health.

DDT, Pesticides, and Human Health

DDT was banned in the United States because of its alleged damage to wildlife. There was little, if any, evidence of damage to humans. Still, one constantly sees reference to DDT as a "known carcinogen" with no reference to who "knows" it and how they know it. When it was banned in 1972 in the United States, "numerous scientists protested . . . that DDT had been widely used during the preceding 25 years with no increase in liver cancer in any of the populations among whom it had sprayed" (Lieberman 1998, 8). Mellanby (1992, 80) says, "The liver cancer deaths in 1944, when DDT was first introduced, numbered 8.4 per 100,000 population. The figure fell to 5.6 in 1972."

Mellanby found this decrease "particularly significant," as longer life spans were putting more people at risk of cancer. He found similar conditions "in other countries where DDT has been extensively used" (80). What was not told in 1972, when it was banned in the United States, or told now, is that using DDT saved millions of human lives. Estimates made by reputable scientists and scientific organizations ran as high as five hundred million lives saved. In 1979, the World Health Organization monitored the use of DDT, failed to find "any possible adverse effects of DDT," and deemed it to be the "safest pesticide used for residual spraying in vector control programs" (Mellanby 1992, 82).[3] "The excellent safety record of DDT, never matched by other insecticides used in antimalarial campaigns, other vector control programmes, and agriculture, is based mainly on its poor absorption through the skin" (WHO 1979, 145).

A study on the relationship between DDT (dichloro diphenyl trichloroethane) and PCBs (polychlorinated biphenyls) with breast cancer that was "larger and better designed than any before it" found "no evidence that exposure to the chemicals DDT and PCB increases the risk of breast cancer" (Kolata 1997; Hunter et al. 1997; BMJ 1997; Krieger et al. 1994; Safe 1997a,b). The study "came as a shock to some advocates for patients," while a spokesperson for an advocacy group maintained that the study was "definitely not the last chapter" (Kolata 1997). It should not in fact have come as a shock to anyone who had followed the debate on the safety of DDT over the last thirty years. Numerous studies on those most exposed to DDT, including those who were dusted with DDT during World War II, those who worked over ten years manufacturing DDT, and those heavily exposed to DDT in the antimalaria spraying campaign, failed to find any adverse health effects on humans (Mellanby 1992, 73–82; A. Smith 2000).

Nearly three years after the previous report, there was another study, the largest such study to date funded by the National Cancer Institute and the National Institute of Environmental Health and Safety, this time carried out by researchers at the Yale Cancer Center. Once again, they "did not find a significant relationship between exposure to PCBs or the pesticide DDE (a metabolite of DDT), and the risk of breast cancer in women" (YCC 2000; Zheng et al. 2000; Bagga et al. 2000; Longnecker et al. 2000).[4] YCC (2000) states, "Zheng and colleagues have also examined the levels of the chemical compounds DDT, DDE, hexachlorobenzene (HCB) or benzene hexachloride (BHC), oxychlordane and trans-nonachlor in the breast adipose tissue of women with and without breast cancer." And they found that there was "no significant difference . . . in the levels of these compounds between breast cancer cases and their matched controls." Tongzhang Zheng, the lead researcher on the project, stated that he was "hopeful that this, our most recent and conclusive study, will put to rest the notion that exposure to DDE causes breast cancer" (YCC 2000). We can share the same hope but not necessarily the same expectation. Issues concerned with women's health are grossly underfunded. It is doubly tragic, then, that so much of this limited funded research is driven to continue what must be seen as a wasted effort in pursuit of mythical causes of cancer, when there are so many other women's health issues crying to be studied.

Over the years, there have been articles in scientific journals that collated these results of DDT studies (Spindler 1983; Coulston 1985, cited in Mellanby 1992, 75; IARC 1987, 1991; Anderson et al. 1996, 746–47; Longnecker, Rogan, and Lucier 1997; Weiderpass et al. 1998). In the United States, liver cancer in humans declined significantly during the period of DDT's use, contrary to animal studies that found malignant tumors in mice in very heavy dose experiments (Mellanby 1992, 80). Though lacking evidence that DDT causes liver damage, we know that many who do not die directly from malaria will have a damaged liver, leading to chronic ill health and earlier death.

Scientific and technological inquiry is ongoing. Though the "last chapter" for any inquiry has not been written and never will be as long as scientists work, we must act on the best available evidence—not the last chapter that will be written but the last chapter that has thus far been written—which in this case is that DDT and PCBs are not shown to cause breast cancer. Is there any doubt that had the DDT/PCB study reached the opposite conclusion, it would have been deemed final and definitive by those who now question its validity? To antitechnologists, there is always an escape hatch, since this was a "study of two chemicals out of 80,000" in the environment. The 80,000 chemicals are guilty until proved innocent. Since no proof of innocence will ever be accepted by the believers, then "chemicals" are guilty a priori. In this misunderstanding of scientific inquiry, a guilty verdict completely closes out inquiry, while an innocent verdict leaves inquiry open, along with the continuing suspicion of guilt. The passive voice in English is useful referring to "known carcinogens" without having to specify who knows it and how they know it.

What is being asked is that science do what, by the very nature of the process, it cannot do. One hundred percent proof of the absolute safety of a chemical or product throughout the lifetime of its use is demanded. Science does not operate in terms of such certainties. It is one thing to argue that following a specified set of tests and examinations of the available data, little or no evidence exists that a substance is harmful to humans. It is another thing to prove with any certainty that no evidence will emerge in the future. The latter is too often demanded. A scientist cannot guarantee that research and experimentation involving animals will produce a desired result, such as a cure for a disease. The experiment would not have to be carried out if we knew the answer in advance with certainty, since it would not then by definition be an experiment. Not knowing the answer in advance is used as an argument against the use of animals in research.

The media heavily publicize every alleged cancer threat of modern technology and science and also tend to give premature claims for each new potential drug or therapy as "a cure for cancer." Despite the media overreaction, there are a number of promising developments. The issues involved are obviously complex, but the statistical data clearly indicate that there has been steady, definite, important, and significant progress made in the diagnosis and treatment of cancer (Altman 1998). From 1990 to 1997, the age-adjusted incidence rate and death rate for all cancers combined, both fell an average of 0.8 percent per year, with the rate of decline for cancer death rates being most rapid after 1995 (Ries et al. 2000; ACS 2000; NCI 2000).

A 1971 National Academy of Sciences report expresses concerns about DDT throughout the study (NAS 1971, 182, 213, 215, 431). But the study praised DDT for its contribution to "the great increase in agricultural productivity, while sparing countless humanity from a host of diseases, most notably, perhaps, scrub typhus and malaria" (432). The report called for a "rule of reason" since

It is estimated, in little more than two decades, DDT has prevented 500 million deaths due to malaria that would have otherwise been inevitable. . . . Abandonment of this valuable insecticide should be undertaken only at such time and in such places as it is evident that the prospective gain to humanity exceeds the consequent losses (432).

The estimate of "500 million deaths" from malaria that DDT prevented may be a typographical error, but whatever the actual number of lives saved from malaria by using DDT, the testimonies of a number of authorities are not exaggerated when they argue, as Whelan does, that "DDT prevented more human death and disease than any other man-made chemical in human history" (Whelan 1993, 100). The *British Medical Journal* called DDT a "miracle chemical" and claimed that it "has been incontrovertibly shown to prevent human illness on a scale hitherto achieved by no other public health measure entailing the use of a chemical" (BMJ 1969).[5] The then president of the National Academy of Sciences, Philip Handler, called DDT, "the greatest chemical that has ever been discovered" (Whelan 1993, 111).

Malaria is considered by some experts as being responsible for causing "more deaths in human populations than any other human disease" in history (Cohen 1989, 41; Burnet and White 1975, 232). It was the leading cause of sickness and death in 1940 just prior to the use of DDT as an insecticide, despite having been reduced or nearly eradicated in many areas, owing to the elimination of breeding grounds for mosquitoes by draining swamps or bringing land under cultivation (Desowitz 1993, 63).

The movement in the United States to ban the use of DDT had an adverse effect on the well-being of the poorer people in the world. According to a study sponsored by the Institute of Medicine, the National Research Council, and the National Academy of Sciences:

> The declining availability and use of DDT was not without its cost in human health but those who promoted the ban were not those who suffered the consequences. . . . The removal of this cheap and effective antimalaria weapon from the U.S. marketplace had a negative impact on malaria control efforts worldwide (Oaks et al. 1991, 44–45).

There are now many alternatives to DDT or pesticides used in conjunction with DDT, but the fact remains for many areas and many uses, that DDT remains the safest, cheapest, and most effective means for protecting humans from malaria and other diseases or causes of death. This is particularly true for indoor spraying for malaria control (Oaks et al. 1991, 35, 132–33). Although mosquitoes in some areas have developed resistance to DDT or similar chlorinated hydrocarbons, in many areas these chemicals are still the best weapon, or an important weapon, in an arsenal of weapons against malaria (Boyce 1998a; Desowitz 1991; Tren 2000). Even when mosquito nets are used to cover one when sleeping, as an alternative to spraying or as an added defense, it has been clearly demonstrated that the nets are most effective when impregnated with pesticide. There are known recent examples where the malaria cases increased in countries after the use of DDT was stopped, and in Ecuador, the number of malaria cases fell when spraying was increased (Boyce 1998a, 19; Roberts et al. 1997, 2000). Many very poor countries such as Mozambique, dependent on external assistance, have been forced to give up spraying with DDT, and deaths from malaria have risen dramatically (Sidley 2000; WHO 2000a; Ashraf 2000; DeGregori 2000b).

If the aim of those who would ban the production in the world of DDT and other similar chemicals is eventually to prevent their use worldwide once existing stocks are exhausted, then the cost of preserving populations of eagles and peregrine falcons would be the increased incidence of malaria. DDT is now manufactured in only two countries, India and China.

There are three hundred to five hundred million clinical cases of malaria each year, with 90 percent of these being in Africa. Of the roughly two and one-half million people who die each year of malaria, most are in Africa, most are children, and most are poor or very poor. It is also estimated that malaria's presence in an area has a 12 percent adverse effect on income (Brundtland 1998; Tren and Roberts 1999; Yamey 2000b; Kmietowicz 2000). In Africa, malaria has been on the increase for several decades and kills more people than all the conflicts and natural disasters combined (PANA 2000). Yet DDT is referred to as the "notorious pesticide" by the World Wide Fund for Nature in a call for a ban on its use in Africa as part of a global ban on DDT (WWF 1998). The WWF makes a number of charges about DDT, including the standard claim that it causes cancer.

One hundred governments, with United Nations officials and nongovernmental organizations, hope to reach a legally binding treaty by 2000, banning the use of DDT and other POPs (persistent organic pollutants) (WWF 1998; WWF 1999a). The twelve organic pollutants on the initial list at the 1999 United Nations Conference on Persistent Organic Pollutants that treaty supporters are trying to control are aldrin, chlordane, DDT, dieldrin, dioxins, endrin, furans, heptachlor, hexachlorobenzene, mirex, PCBs, and toxaphene. The global ban on DDT may or may not be good for wildlife, but it will be devastating for humans, particularly for children in Africa (Lancet 2000f).

The WWF in reports and press releases is now claiming that the World Health Organization is using obsolete 1993 scientific data to support the continued use of DDT in malaria control programs (WWF 1999c, d). At least one very reputable source uncritically repeated these allegations as fact (Kirby 1999). Yet the WWF fails to cite any of the peer-reviewed studies that I cite, which find no evidence that DDT is harmful to humans. The more recent study (1995) that WWF claims makes WHO data obsolete was criticized at the time of its release (Siddle 2000; Gladen et al. 1995). As I note below, eight attempts to replicate it have failed (Attaran and Maharaj 2000). Checking the WHO website, one finds WHO concerned about POPs but arguing that DDT "remains a valuable tool" for disease-vector control, particularly for malaria, however they oppose using DDT for agriculture (WHO 1999a).[6]

Opposition to the manufacture and use of DDT goes back about three decades, which raises the question as to whether the opponents now recognize that there was in fact data prior to 1993 that supported the continued use of DDT in malaria control. If they were wrong then, might they also be wrong now? The language in the WWF (1999b) "Issue Brief" is anything but definitive on the subject: "It is, therefore, not surprising that humans seem to be suffering increasingly from the same health problems reported in laboratory animals and in wildlife exposed to one or more of the dozen POPs." Later the report adds: "Although the pattern of evidence is highly suggestive, it is virtually impossible to answer questions about the impact of these persistent chemicals on human health directly or definitively."

Other anti-DDT reports are also less than definitive in establishing causality for harmful human effects. On the website of the Environmental Health Center (National Safety Council), I found: "DDT is listed in the National Toxicology Program's Fifth Annual Report on Carcinogens as a 'substance which may reasonably be anticipated to be carcinogen.' The EPA categorizes DDT, DDE, and DDD as probable human carcinogens" (EW 1997).

"Reasonably be anticipated to be carcinogen" or "probable human carcinogens" should be weighed against the known benefits that we have received from these chemicals. On the basis of "seem to be" or "highly suggestive," the WWF and its allies would have us ban DDT, which is still the most effective protection in many areas against malaria, as well as dengue fever and leishmaniasis.

The WWF does recognize that malaria remains a serious threat in over ninety countries where roughly 2.5 billion of the world's poorest people live. "It is a leading cause of illness and death in the developing world, contributing to approximately 3 million deaths and up to 500 million acute clinical cases every year" (WWF 1999d; Croft 2000). Most of these deaths occur in Africa, and children under five are the major victims: "malaria kills four children per minute or 5,000 per day" (WWF 1999d). Tragically, malaria-bearing mosquitoes appear to have a preference for pregnant women (BMJ 2000b). Currently, there is a major increase in chloroquine-resistant malaria throughout southern Africa (Nduru 1999),[7] and malaria is now appearing again in colder countries such as Azerbaijan and Tajikistan (IFRCRCS 2000a,b). Unanswered by DDT's opponents is why people are still using DDT (or trying to do so) if it is ineffective or if there are equally effective alternatives? It would not have to be banned if it were not getting the job done.

Dr. Charles Wurster, chief scientist of a leading environmental group, Environmental Defense Fund (now called simply Environmental Defense), when told that banning DDT would lead to millions of deaths worldwide, responded that there were too many people and "this is a good a way to get rid of them as any." The remark led to the resignation from the organization of Victor Yannacone, the attorney who had won the first successful case to ban DDT (Whelan 1993, 67).

Similar sentiments have been expressed in discussions about the tsetse fly (*Glossina* sp.), which infects Africans with the parasitic protozoan disease, trypanosomiasis (sleeping sickness), keeping them from inhabiting certain areas by killing their livestock. The tsetse fly has been called "the best game warden in Africa," which shows the same callous disregard for the well-being of Africans that I discussed above (Adams and McShane 1996, 49). In developed areas, such as New York City, it only takes a small number of people, three in this instance, to die of a mosquitoborne illness, West Nile virus, for a massive spraying operation to be undertaken. This spraying campaign in New York was taking place at the same time that there were discussions at the UN about phasing out the POPs (Tierney 1999). There are concerns about the return of malaria to Europe and North America (Gilhooly 2000).

A British scientist, Dr. Norman Moore, was the first person to advance the claim in the 1950s that DDT was causing the decline in the population of eagles. Wakeford (1991, 34) gives Moore credit as the person who "initiated the research that showed how pesticides damage wildlife and framed the laws that brought these chemicals under control." Moore himself was sprayed with DDT when liberated from a prisoner of war camp in 1945, and has suffered no adverse effect over fifty years later. Moore refuses to condemn all uses of pesticides, pointing out that they have increased food production and saved millions of people from insect-borne diseases. He adds, "If I were living in a hut in Africa, I would rather have a trace of DDT in my body than die of malaria" (37).

After World War I, two to three million people died of typhus (Borg 1999, 776). There was every likelihood, if not virtual certainty, that much larger numbers would have died of the disease due to the widespread devastation. In an effort to avert tragedy, the Allied forces began a delousing program in Naples in December 1943. DDT mixed with an inert powder was blown into clothes in a matter of seconds and generally succeeded in killing all body lice in an hour or two (Burnet and White 1975, 161). They note, "Some tons of DDT dusting powder and the necessary blowers were available and for the first time in human history a menacing typhus epidemic was abruptly terminated." Like the "curious incident" of the dog that didn't bark in Conan Doyle's Silver Blaze, the story of the last half century might be told in terms of the epidemics that didn't happen.

When the life cycle of malaria was discovered in 1898, a variety of control methods was initiated. Efforts were made to control urban malaria by draining swamps, by using many of the very harsh, highly toxic substances employed at the time to control insects in agriculture, and by using window screens and nets to keep mosquitoes out of sleeping quarters. Not until the 1940s was rural and village malaria successfully attacked, and Jukes (1974, 6) attributes the success to DDT. As I noted in Chapter 2, DDT was also safely used to eradicate onchocerciasis in Kenya in the 1940s and in Uganda in the 1960s, and it was done in Kenya at the low price of $1.82 to $3.35 per square kilometer. From the mid-1940s to the early 1970s, the use of DDT in onchocerciasis control programs spared countless millions from this horrible disease.

In many areas of developed countries where malaria was once endemic, this combination of environmental modification and DDT (and other pesticide use) has completely eliminated malaria. But, it is also true that travel, population movements, and opening new areas for habitation have had the opposite effect by spreading the disease (Martens and Hall 2000). Environmental methods of control are difficult and prohibitively expensive for rural areas in tropical countries (WHO 1971, 181). The swamps that would have once been drained are now wetlands with their unique flora and fauna that environmentalists seek to preserve. The environmentalists can't have it both ways, preserving the breeding areas for malarial mosquitoes and opposing a critical component of the most effective means of controlling them.

The authors of an article titled "Balancing Risks on the Backs of the Poor" call the attempt to ban DDT "eco-colonialism that can impoverish no less than the imperial colonialism of the past did" and is once "again to embrace indifference, and the pursuit of environmental goals on the backs of the world's sickest and poorest." Further, the first study to find an elevated risk of breast cancer from exposure to DDT "has now failed to be replicated at least 8 times" with some studies even finding "significantly" reduced risk with similar findings for "multiple myeloma, hepatic cancer and non-Hodgkin lymphoma" (Attaran et al. 2000, 731, 730).[8] Attaran and Maharaj (2000) added that although "hundreds of millions (and perhaps billions) of people have been exposed to elevated concentrations of DDT . . . the literature does not contain *even one* peer-reviewed, independently replicated study linking DDT exposure to *any* adverse health outcome."

To the eight studies failing to confirm a link between DDT and breast cancer cited by Attaran et al. (2000), a ninth can now be added by Stellman et al. (2000). Ironically this study "was funded as part of the Long Island Breast Cancer Study Project which was created in 1993 by Congress after lobbying by local activists" and was sponsored by the National Cancer Institute (Fagin 2000). It is legitimate and not merely rhetorical to ask whether the "activists" could have done more to improve the health of women by using their lobbying skills to obtain funds for research on women's health that is less politically correct but has more promise of an outcome that identifies a cause of a malady and its possible cure.

As a result of a rising tide of opposition by African and other developing country leaders and their governments and by public health workers and professionals around the world, and by the actions by concerned citizens and professionals such as Attaran, in the fifth and final round of negotiations (Johannesburg, South Africa, December 2000) on a treaty to ban the POPs (persistent organic pollutants), DDT for use in disease-vector control was "exempted" from the ban and placed in a special category. DDT use will continue to be primarily for malaria control but it is also essential for other vectorborne diseases such as dengue fever. Use of DDT can continue until a safe alternative that is as cheap and effective as DDT can be found, with the burden to do so being on those who wish to ban DDT and not on those who wish to use it to protect their population, particularly their children, from malaria and other debilitating and death-dealing vectorborne diseases.

A New Mosquito

Another malaria weapon, a genetically transformed mosquito, soon will be used to investigate molecular aspects of malaria parasite transmission. Coates (2000) predicts that it could lead to new control mechanisms for the disease (900). It would involve genetically engineering a strain of mosquito that could not transmit the pathogen, and replacing the wild population with this "innocuous strain" (900; Catteruccia et al. 2000). Those involved in this research are very much aware of the complex issues involved in releasing a genetically transformed mosquito (Bradbury 2000). We all should be concerned about possible unintended adverse outcomes of releasing a transgenic mosquito into the environment, but doing nothing means more deaths from malaria. Actions that ban some pesticides or that make it needlessly difficult to get others approved often lead to greater use of pesticides, not less. Many of the so-called natural pesticides (made from plants) are often used in much higher doses than chemical pesticides. Pyrethrin (from the chrysanthemum family) is one of the most widely used plant-derived pesticides. It has been used for several centuries and was called Persian Insect Powder (Whorton 1974, 15–17).

The trend in modern agriculture is toward the use of lower dose pesticides that are more potent but more targeting, less persistent, and less toxic per hectare, reducing soil contamination (Gianessi 1993). The addition of "new active ingredients" has given the user a wider range of choices. "Improved formulation and application techniques" have meant a fall in the typical application rates of herbicides from 3,000 grams per hectare in 1966 to 100 in 1987, with insecticides falling from 2,500 grams per hectare in 1965 to 20 in 1982, and fungicides from 1,200 in 1961 to 100 in 1991. In 1945 DDT was applied at a rate of about 2 kilograms per hectare. Similar pest control can be achieved through today's pesticides at the rate of 0.1 kilogram per hectare and even 0.05 kilogram per hectare (Waage 1995; Yudelman, Ratta, and Nygaard 1998, 9, 11).

Disease-resistant Plants

Many of the green revolution crops were bred to be disease resist-ant, and therefore many of them require a lower use of pesticides to get the same crop protection. In addition, when a crop in a region is hit by a pest infestation, new varieties resistant to that infestation are often in-troduced. As we have seen, plants have been genetically modified specifically to be used with herbicides that provide more efficient and effective crop protection. Yet there is a persistent, unqualified, widely and fervently believed antitechnology mythology that claims that green revolution crops are less disease resistant.

At the very beginning of the green revolution's research on wheat in Mexico, research was conducted on developing rust and smut-resistant varieties of wheat as a priority to higher yielding varieties. The reason-ing of the Mexican government was that there was little purpose in in-creasing output if it was going to be destroyed by disease. The control of stem rust by resistant varieties has made an important contribution to wheat output in North America (Oerke 1994, 185). Genetic modifica-tion is now creating wheat with even greater resistance to "stinking smut" (Clausen et al. 2000). Even if the modern varieties experience larger crop loss and require greater use of pesticides, the vastly in-creased yield means that there is still much more food to feed people.

Most studies of pesticide use over the last couple of decades indi-cate that farmers get a return of anywhere from four to six dollars in increased crop yield for each dollar invested in chemical pesticides (Farah 1994, 1). The lower yields without pesticides, cited above, re-sult in a higher cost per unit output. The estimates are that corn (maize) would cost 27 percent more per unit of output without pesticides and 61 percent without any chemicals, while for wheat the cost increases would be 33 and 50 percent per unit of output (Knutson et al. 1990, 5).

Decreased output and increased costs mean higher food costs for the consumer. Using the above production and cost estimates, the result would likely be a 12 percent increase in food costs for middle-income consumers. For the 20 percent of the U.S. households with the lowest incomes, the no-chemical scenario would mean an increase in the per-centage of their income spent on food from 38 percent to 44 percent

(Knutson et al. 1990, 42, 44). It would also mean a sizeable reduction in U.S. food exports, reducing global food availability and increasing food costs for the poor. The no-chemical scenario also assumes an increase in land brought back under cultivation, which would have adverse environmental implications and a reduction of food stocks, thereby increasing vulnerability to output declines due to weather or other factors, and thus decreasing food security.

Increased Yield, Increased Problems

Modern varieties of food crops (rice, for example) are frequently bred to have shorter growing seasons, allowing more plantings per year. In the tropics, many plots of land are continuously cropped throughout the entire year. The modern varieties provide a significantly increased yield per plant and per crop, and an even greater increase per year. Using fertilizer for nutrient, crops can be planted more densely. All of which means that there is a vastly increased amount of food in the fields, not only for humans but also for those who compete with humans for the crop's nutrient. With little break in crop production, there is less time for these competitors to die out, or at least be dramatically reduced in numbers. It is understandable, then, that protecting the crop becomes more difficult and that proportionate losses may equal, or at times exceed, those for traditional varieties. Generally, a "crop's susceptibility to attack by pathogens and animal pests increases with yield" (Oerke et al. 1994, 765). However, this simply does not negate the resistance that has been so effectively bred into many of our food crops.

The more intensive the system of cultivation, the larger are the potential losses due to harmful organisms. Even if traditional varieties could generate this high level of output, there would still be problems of pest control. There is no evidence that the traditional varieties could do so, and the point is moot anyway, since they cannot produce the required output. Even if pesticides are used, there are many other forces that can "stress" a crop and reduce or destroy output, such as water (too much or too little), heat, or cold. In addition to being bred for disease resistance, many green revolution crops were bred to be better able to adapt to a wide variety of stress-causing conditions.

To maximize food production (including food preferences), income, or some combination of the two, farmers, farming cooperatives, and agencies advising farmers have a variety of agricultural strategies they can pursue. They can grow more disease-resistant varieties and economize on pesticide use but possibly have a lower total and net output. They can rotate crops and use other strategies that limit pest damage, but again, there is very likely to be total output loss. Sometimes a higher valued crop (aromatic rice such as basmati) can be grown, producing a lower output and sold at a price that allows twice as much regular rice to be bought with the income. Sometimes the farmer has the best of all worlds in being able to opt for a variety of the staple crop that is resistant to the broadest array of possible stress conditions and still is the highest yielding. A variety might be chosen for the length of growing season, which may have a lower yield per crop but allow more crops and therefore more output per year. Some farmers will plant a traditional variety when an input or inputs, such as water in monsoon climates, are not certain, and a high yielding variety for the irrigated crop when the water is more secure, thus warranting cash outlays for other inputs such as fertilizers and pesticides, to reap the full benefit of the higher yielding variety.

Whatever strategy the farmer uses, some form of crop protection will be required. Predominantly biological strategies of crop protection will be more labor-intensive and may require cash outlays to pay someone to monitor (called scouting) insect infestation. The farmer can choose a variety, that because of density of output, cropping pattern, or genetic characteristic, requires chemical defense to achieve full benefit. This will be an economic decision and will be done only if the cost of treating the crop with chemicals is less than the market value of the expected crop loss without pesticide use, and if the net output minus all input costs is greater than alternative varieties, cropping patterns, or crop protection (Oerke 1994, 39).

Benefits of the Green Revolution

No matter how one compares losses, the agricultural technology of the green revolution has made more food available to far more people than at any time in human history. From 1945 to 1989, many important crops, including some that are heavy users of pesticides, such as corn (maize), more than doubled their yield per acre. The accompanying increase in insect loss was actually a decrease in loss per unit of output and still resulted in a massive increase in food availability.

The fact that it took a tenfold increase in pesticide use to achieve the higher output is a cause for concern. A diverse array of pesticides and nonchemical strategies—IPM for example—was created in the continuing effort to produce enough food to feed the world and preserve the environment.

Biotechnology offers increased possibilities of breeding more disease-resistant crop varieties. Tragically, some of those who claim to be most opposed to pesticide use also oppose the use of biotechnology in plant breeding. Because of the direct relationship between yield and the need for chemical crop protection, antitechnology beliefs and actions will make it increasingly difficult to feed the world's growing population (Oerke et al. 1994, 765). In some cases, groups like Greenpeace use totally illegal means to further their antitechnology, antibiotechnology agenda (Miller 1997, 11).

"Precision farming" involves the use of computers on tractors combined with GPS accessing satellites to measure precisely how much seed, fertilizer, and pesticide are placed on each area of a farm and, later, how much is harvested from each area. Computers with IPM expert systems software and inexpensive equipment, such as rain gages and barometers, are being used with a procedure called scouting to decide when to use pesticides (and what kind) and when not to use them. Farmers generally do not wish to use inputs, such as seeds, fertilizer, and pesticides needlessly, since they cost money. Precision farming allows the farmer to get the largest output per unit of input and efficiently substitutes technology in the form of knowledge for inputs, such as fertilizer and pesticides.

This package of technologies is usable only by large farms, primarily in developed countries. The task we face is to adapt and adopt what is learned from this endeavor in ways that are affordable for poorer farmers in developing areas. The techniques of precision farming that minimize pesticide use are generally beyond the capability of farmers in poor countries. In addition, the use of the safer, less-contaminating pesticides largely occurs in developed countries, while the earlier generations of pesticides tend to be used in developing countries (Waage 1995, cited in Yudelman, Ratta, and Nygaard 1998, 9; Oerke 1994, 752).

Domestic laws that prevent the import of produce carrying residue of chemicals banned in the United States force growers to use pesticides that are more toxic to the farmworkers and more expensive (Perfecto 1992, 189; Perfecto is a critic of most uses of pesticides). Perfecto (1992) says, "To ensure the entry of vegetables into the U.S.,

Mexican farmers switched from highly persistent organochlorines, which were banned or restricted in the U.S., to the less persistent organophosphates" (189). As would be expected, the shift to pesticides that were more volatile and toxic was harmful to those who worked in the fields. In Perfecto's words, "Organophosphate insecticides are more acutely toxic than the compounds that they have replaced. They were recommended by the EPA because they degraded faster and wouldn't persist in the environment or food" (189).

The same attempt to lower pesticide residue on domestically produced food in the United States means that "moderately toxic but persistent" pesticides have been replaced by organophosphate pesticides that are "less persistent but acutely toxic to workers" (Gray and Graham 1997, 189). The consumers' presumed gain has been at the expense of agricultural workers.

Child Labor

Misguided U.S. legislation against the importation of products produced by child labor in order to "protect" the children from exploitation and promote their education has had a devastating impact in Bangladesh on the lives of those for whom it was designed to protect (Rahman, Khanam, and Absar 1999; Quddus 1999).

Not only do consumers in developed countries (and the legislators that they support) show little concern for farmworkers in developing countries, they show little concern for farmworkers in their own country, even when these workers are children. An excellent series of articles by the Associated Press at the end of 1997 showed how poorly child labor laws are enforced in the United States, particularly in agriculture (Mendoza 1997). In the United States, the Food Quality Protection Act of 1996 requires the Environmental Protection Agency to consider children, who are most vulnerable to such health effects, when it sets limits for pesticides in food. The law, passed unanimously by the 104th Congress, was so strict that even the environmentalists who lobbied for it could hardly believe it (Crenson 1997). The Food Quality Protection Act "specifically forbids" the EPA from "considering occupational exposures, which are known to be much higher than those generally found in food and water. The act goes out of its way to exclude the children and adults who pick America's produce" (Crenson 1997). This practice of excluding children in agriculture from laws protecting children is of long standing and continues to the present (HRW 2000; Whitman 2000).

Testing Pesticides on People

It is hard for some not to believe that the environmental movement is more concerned with middle-class consumers than with lower-class workers. The law was an "explicit, purposeful exception" according to Ralph Lightstone of California Rural Legal Assistance. "They carved farm workers out," he said (Crenson 1997). Subsequent attempts by legal and administrative action to protect the children in the fields from possibly excessive chemical exposure have clearly not been successful (GAO 2000b).

The law in its very strictness has had the unintended consequence of increasing the testing of pesticides on human beings. The tough standards embodied in the act make it almost impossible to demonstrate product safety by animal testing alone. Consequently, companies can only demonstrate a product's safety for humans by testing it on humans.

There are other laws that regulate worker access to fields when they are sprayed, but pesticide levels in the farmworkers' blood indicates that the laws are either not being enforced or need to be strengthened. Similarly, in the United States, we have laws to protect the industrial workplace, but any failure in enforcement or need for improvement does not draw the attention of the media or the environmental movement. To the extent that the adage, "dose makes the poison" is true, we are legislating to protect those who do not need it and ignoring those who do.

Health Risks Near Pesticide Plants

So far, there is no evidence that the manufacture of pesticides provides any health risk for those who live near these facilities or (as we noted for DDT) for those who work there. Researchers from the University of Birmingham in the United Kingdom studied over 22,000 childhood deaths before the age of sixteen "that occurred over a period of 27 years between 1953 and 1980, and found that "children born near industrial sites that use high-temperature furnaces or discharge volatile organic compounds are more likely to develop cancers such as leukemia." Looking at environmental factors, they "found significant geographical clusters of cases near industrial sites." They also found that "large hospitals with big incinerators could be a cause of childhood cancer," whereas they confirmed that nuclear installations and pesticide factories did not present a risk (BBC 1998b; Knox and Gilman 1998, 720–21).

Sickness and Death Due to Pesticides

The data on the health costs for pesticide use have been hotly debated over the years. Critics have charged that the estimates for sickness and death due to pesticide use have been greatly inflated. For example, the figures for pesticide poisonings in the United States tabulated all inquiries made to poison centers on the dangers from spilling or swallowing pesticides without indicating what the actual health outcome was. For the global figures, estimates of deaths from pesticide routinely include those where the pesticide was used to commit suicide. It should be noted that arsenic, the main pesticide that preceded modern chemical pesticides, was sometimes used to commit homicide and suicide (Whorton 1974, 38). Whatever one's view of the dangers or lack thereof from pesticide residue on food (discussed elsewhere in this book), the level of exposure for farmworkers can be a problem. Even if one argues that most estimates exaggerate the sickness and deaths, one characteristic pervades virtually all estimates, namely, illness and death associated with pesticide use is vastly greater in poor countries than in developed countries. Farah (1994) notes, "Studies suggest that up to the mid-80s, about half of all pesticide poisonings and 80 percent of pesticide-related deaths occurred in developing countries, even though this is where only 20 percent of pesticides were used" (Farah 1994, 4).

Farah found that in Sri Lanka, with a population of fourteen million, "2.5 times as many people suffered pesticide poisoning and 5 times as many people died in 1978" compared to the United States with a larger population, where one-third of all pesticides are used (Farah 1994, 4). This would seem to indicate that it is not pesticide use that is dangerous per se but the way in which pesticides are used and what they are used for, as Sri Lanka is a country where there have been reports on the use of pesticides to commit suicide. Clearly, continued attention to usage is important if the inquiry is rational and scientific. Prudent use of pesticides can benefit countries such as Sri Lanka.

Importing and Exporting Pesticides

There is an FAO-brokered agreement that requires notification and explanation to importing countries (called "Prior Informed Consent") if the pesticide has been banned in a number of countries (FAO 1990; Farah 1994, 6–7). A similar agreement also involving FAO now exists in treaty form (Connole 1998; Nabakwe 1998). In addition, for U.S. firms there is a series of domestic laws that require "foreign governments to be notified before consumer goods banned in the United States" can be exported. Pesticide exporters are required "to obtain a statement from the foreign purchaser which acknowledged that the pesticide could not be used in the United States" (Vogel 1997, 197).

The point is not whether a pesticide is banned for use in the United States, but whether all participants in the process—seller, buyer, and user—are reasonably informed of the likely beneficial and adverse consequences of its use. There is a critical and ongoing need for *knowledge* among the buyers and users of a product and among *those who seek to restrict its use*. What may be an action motivated by high ethical principles in a developed country can, as a consequence, worsen the condition of the poor in developing countries.

Notes

1. For an excellent discussion of the enormous complexity of determining what is or is not toxic in the environment, see the essays in Foster, Bernstein, and Huber (1993).

2. On the hazards of biomass fuels, see Koning, Smith, and Last (1985); Kamat (1998), 57; and Barnes et al. (1997).

3. See WHO (1979).

4. On PCBs and thyroid hormone, see Longnecker et al. (2000).

5. Quoted in Whelan (1993), 103.

6. On the economic costs to Africa of malaria, see Boseley (2000) and Voelker (2000b).

7. Chloroquine was for many years the standard drug taken for protection against malaria.

8. The eight studies referred to by Attaran et al. (2000) are Krieger et al. (1994), Schecter et al. (1997), Hunter et al. (1997), Lopez-Carillo et al. (1997), Maysich et al. (1998), Zheng et al. (1999), van't Veer et al. (1997), and Helzlsouer et al. (1999).

CHAPTER 7

The Nazis and the "Natural"

I n looking at romantic technophobic belief systems, I can draw similarities to many ideas advocated by leading Nazis in the 1930s. There are far too many to be mere happenstance. I will try to be careful in highlighting the similarities. I agree fully with Proctor when he cites the ethicist Arthur Caplan on the dangers characterizing anything as Nazi lest we end up "diminishing the genuine extremity of the Nazi experience" (Proctor 1999, 271; Drotman 2000). Just because I argue that the Nazis were probably the first antitechnology postmodernists does not mean that one can reverse the argument and claim, therefore, postmodernism is a contemporary version of the Third Reich, with all the horror this implies. I am hesitant to make such connections because the evil of the Nazis is so great that it is difficult to think only in terms of a limited set of comparable ideas. However, many of the beliefs are so strikingly similar between the Nazis and animal rights activists, environmentalists, and postmodernists that I cannot totally ignore the issue.

Similarities between Nazis and Contemporary Movements

The list of similarities between contemporary movements and the Nazis is in fact quite substantial. At the top of the list is the opposition to rational, scientific inquiry, which is deemed to be "reductionist," while preferring synthesis over analysis. Following from this is a romantic view of nature and what is natural; organic agriculture (biodynamics—opposition to manufactured fertilizer and pesticides); vegetarianism; animal rights; a view of conservation involving the forced removal of people; holistic healing and medicine; and the "blood and soil" garden (opposition to "alien species"). Those activists who opposed genetically modified crops by "rooting out genetic pollution" in the fields are unfortunately describing their actions in language that is all too reminiscent of the 1930s, though I would not wish to trivialize the latter by comparing them in any other way. Yet, underlying all of these similarities is a shared sense of the "purity" of their thoughts and actions and a sense that they are acting in defense of self and defense of higher moral principles.

Danger in Response to Higher Calling

It is this sense of purity, and a belief that they are responding to a higher calling, that I find to be dangerous, whether to preserve the purity of the *volk* or to save planet earth from contamination from genetically modified crops. These beliefs validate taking forceful actions that interfere with the rights of others. "No compromise in defense of mother earth" means that others are prejudged to be guilty, so that any action taken that harms them is simply punishment for their sins. An apocalyptic view that the earth is threatened with imminent destruction absolves one from having to understand the consequences of one's actions, since the good of saving the planet cleanses any wrong that one may do in saving it. A problem in fully criticizing these actions is that the verifiable damage actually done may be trivially small compared to the larger consequences of the unverifiable harm that follows from it. No one can say for sure that someone whose research was disrupted by animal rights activists would have, without the disruption, found a cure for a disease that debilitates tens of millions of people or that destroying research in agricultural biotechnology might have prevented advances in agriculture that would feed hundreds of millions. Science and technology do not deal in such certainties, but purists do. We can demand that these questions of harm of major proportions be part of the discourse and debate on these issues.

Vegetarianism and Nazi Ideology

Vegetarianism and commitment to organic agriculture were a component in Nazi ideology. According to Arluke and Sax (1992), Hitler
and other elite Nazis became vegetarians in emulation of Wagner.
"Rudolf Hess's food had to contain 'biologically dynamic ingredients'" (17). "As sacred things, society was not to violate animals by
killing them, either for sport or for food. . . . Hunting became a symbol of the civilization left behind; meat eating . . . a symbol of the decay of other civilizations" (17).

"Vegetarianism became the symbol of the new, pure civilization that
was to be Germany's future" (17). As I noted above on organic agriculture, many Nazis were exponents of the philosophy of Rudolf
Steiner and "bio-dynamic agriculture." Synthetic fertilizers were "rejected because they were alien to the land, man-made, or at least, man-
extracted, and dead," concepts and terminology still prevalent today
among organic agriculture advocates (Bramwell 1989, 20; Steiner
1958; Ferguson 1997). Robert Pois and Anna Bramwell extensively in-
vestigated Germany in the 1920s and 1930s and the kinship of the organic or biodynamics enthusiasts with certain elements of the Nazi
movement (Pois 1986; Bramwell 1989; Dominick 1992; Cooter 2000).
Included was "the preference of early Nazi health policy for holistic
medicine and natural healing over decadent 'jewish' scientific medicine" (Aly, Chroust, and Pross 1994, 9; Proctor 1988). There were also
strong elements of support for animal rights and a love of animals and
zoos (Arluke and Sax 1992; Foner 1993). In the 1930s, "the University
of Berlin . . . instituted 86 courses connected to veterinary sciences as
applied to humans" (Arluke and Sax 1992, 23). "The Nazis also sought
to protect wildlife" (Arluke and Sax 1992, 8). With the Nazis and some
contemporary animal rights advocates, attempts to elevate animals often end up degrading people.

Wildlife Conservation and Nazi Purism

Post–World War II ideas on wildlife conservation in Africa had origins other than the colonial countries. Bernhard Grzimek, zoology cu-
rator at the Frankfurt zoo under Hitler, was famous for the book and
Oscar-winning film *Serengeti Shall Not Die* and honored as the father
of modern African conservation. Conservationist George Monbiot

quotes him, arguing that "a National Park must remain a primordial wilderness to be effective, no men, not even native ones, should live inside its borders" (Monbiot 1999; Grzimek and Grzimek 1961). Nazi purism in its purist form! Grzimek's ideas were in line with already established British colonial wildlife conservation policies in the Serengeti.

The Nazi prescription for the "conservation of native peoples in their native habitats," described Robert Proctor as being "conceived as part of a larger design to save the world's vanishing flora and fauna," seems benign but in practice, it is highly exploitative, forcing indigenous populations to live traditionally as defined not by them but by outsiders (Proctor 1999, 266).

Natural Gardens

The number of beliefs and practices that are shared by contemporary nature zealots, the Nazis, and other German nativist movements, extends even to the relatively trivial (Biehl and Staudenmaier 2000). Currently in the United States, there is a "natural garden" movement. The "new American garden" is not new. The German nativists and, later, the Nazis also sought to give the German people a "blood-and-soil-rooted" garden, free of alien species. The Nazis "thought with their blood" and acted in terms of it, namely "racial hygiene" (Gay 1968, 82; Decker 2000). One Nazi scientist understood ecology as a "view of the world, in which everything is related to everything else," and equated ecology with the "doctrine of blood and soil," interpreting ecology as serving the interests of the *volk*, just as contemporary environmentalists seek to create a "political ecology" to serve as the "guiding science of postmodernism." Both then and now, science as practiced was, and is, criticized as being "reductionist" (Deichmann 2000, 739).

"Rules of the Design of Landscape" from a group under the aegis of Heinrich Himmler formulated a "close-to-nature" style that required the "exclusive use of Native species" (Pollan 1994, 54). It was important enough to the Nazis that in 1942, during World War II, "a team of Saxon botanists working for the Central Office of Vegetative Mapping embarked on 'a war of extermination' against *Impatiens parviflora,* a small woodland flower regarded as alien" (Pollan 1994, 52). Himmler also had a grand design of creating an "enormous wildlife preserve" in Poland. Himmler had "organic honey," raised at Dachau,

and herbal gardens to provision the German army. Prior to the World
War II, the Nazi Party sought to corner the mineral water market (Drotman 2000).

Animal Protection

One of the first groups that tried to blur the distinction between humans and animals were the Nazis. The "new Germans were to be part
animal, renouncing a certain side of their humanity. The compassion
normally reserved for humans was to be redirected toward animals,
and the cold aggressiveness of animals became the model. Animals
were to be identified with, and compassion toward animals, rather than
humans, was to be encouraged, if not required. This was, in fact, part
of the intent of animal protection laws" (Arluke and Sax 1992, 10–11).
To the extent that the Nazis did distinguish between humans and animals, they "allied themselves with animals, since both were portrayed
as victims of 'oppressors' such as Jews" (Arluke and Sax 1992, 7). "It
is well known that the Nazis treated human beings with extreme cru-
elty but it is less widely recognized that the Nazis also took some pains
to develop and pass extensive animal protection laws" (Arluke and Sax
1992, 6). "In addition to the laws against vivisection and kosher
slaughter, scores of additional legal measures regulating treatment of
animals were enacted from 1933 through 1943" (Arluke and Sax 1992,
7–8). "The fact that animals were to be protected for their own sake,
rather than for their relationship to humanity, was described as a new
legal concept" (Arluke and Sax 1992, 8). In elevating animals, the
Nazis ended up degrading human beings.

Romanticizing the Past

Idealizing or romanticizing the past not only is wrong, but also can
be dangerous. Robert Pois shows that "the myth of the large morally
pristine, pre-industrial family" in Germany in the 1930s not only was
believed, but also was acted upon. "A fetishistic cultural pessimism necessitated the creation of a history-defying totemistic past." Any group,
such as the Jews, who could not fit into the myth were a threat to it
(Pois 1986, 142; Moore 1978; Rhodes 1980, 179–80; Decker 2000).
The Nazis "exalted synthesis against analysis, unity and wholeness
against disintegration and atomism, *Volk* legend against scientific truth.

Life had an organic unity . . . the invisible force that makes the whole more than the sum of its parts" (Arluke and Sax 1992, 12).

In *Conversations with Hitler,* Hermann Rauschning (1940) quotes Hitler:

> We stand at the end of the Age of Reason. A new era of the magical explanation of the world is rising. There is no truth, in the scientific sense. That which is the crisis of science is nothing more than that the gentlemen are beginning to see on their own how they have gotten onto the wrong track with their objectivity.[1]

This quote is in an article that argues that many of these same views hostile to reason have become pervasive among the educated population in the United States and Europe. Slakey adds that currently, the "elegant and modern fashion in thought" is to "reject objective reason and question the value of science, embrace subjectivity and trust your feelings." Slakey adds that "according to anti-science intellectuals, contemporary science has taken us maddeningly close to the edge of the abyss and it will take us over the edge if we are not careful." John Lukacs, contemporary postmodernist guru, is quoted saying that it is not given to "humans to explain everything" (Slakey 1993, 49). Slakey counters that it is the "abandonment of objectivity and not the practice of science that leads to the abyss" (49). A reasonable case could be made that the Nazis were the first postmodernists, which is ironic, because many postmodernist writers, particularly among the French, have tried to argue that the Nazis were the logical outcome of modernity (DeGregori 2000c).[2]

The Path of Reason

This does not mean and is clearly not implied by the authors cited, or by me, that postmodernists or contemporary romantics or romanticism will lead to totalitarianism, with the horrors that it brought in the twentieth century. Those of us who favor the path of reason do not share the apocalyptic view of postmodernism. However, there is still reasonable concern that those who speak of "reenchantment" of the world or offer an idyllic return to a decentralized community living in harmony with nature may be trying to follow a pathway with consequences that not even the postmodernists desire.

It was modernism in the Weimar Republic, particularly in Berlin, that the Nazis were rebelling against. The Berlin of the Weimar was truly one of the greatest cities that the world has ever known (Hall 1998). Berlin was a beehive of creative activity. "Here the formal principles of modernism were applied realistically" to an array of "socially oriented tasks" and to the "communication of intelligible statements and messages" (Willett 1984, 13; Hall 1998, 243). Gay (1968) describes the reaction to the Weimar Republic as a "revolt against reason" (81).

Peter Gay argues that a "hunger for wholeness" was "a great regression" that was "born from a great fear: the fear of modernity." The "wholeness" sought by those rebelling against the modernism of Weimar culture in Germany "was awash with hate" (96). The "desperate need for roots" unfortunately also involved a "vicious repudiation of reason" (Hall 1998, 274). Gay (1968) says, "Most Germans did not want to learn . . . the lesson of Bacon and Descartes and the Enlightenment: that one must confront the world and dominate it, that the cure for the ills of modernity is more modernity and the right kind of modernity" (101).

Clearly not all repudiation of reason will give rise to forces of hate; however, the repudiation of reason carries a price, and it is one of sufficient concern that the rose-tinted glasses of romanticism cannot hide. Further, for the antitechnology postmodernists and others who hold to this constellation of ideas, one would hope that this significant overlap in beliefs with the Nazis would make them uneasy and force them to do what we all should do, that is, reexamine and rethink our belief systems. Peter Gay is as right about our own time as he was about the Weimar/Nazi period in arguing for "an appreciation of the true freedom that comes with rationality and discipline" (142).

Notes

1. Quoted in Slakey (1993), 50.
2. On the connection between the Nazi ideologists and early leading lights of postmodernism, see Windschuttle (1997), 178–83.

CHAPTER 8

Technology, Food Production, and Drinking Water

I t is hard to imagine anyone opposing the green revolution. The increase in yields from the green revolution in rice alone has produced enough to feed one billion people (DeGregori 1987b). Growth in yields accounts for 92 percent of the increase in world cereal production since 1960 (Frisvold, Sullivan, and Raneses 1999). The world average for grain yields per hectare rose from 1.1 tons in 1950 to 2.9 tons in 1992 (Conko and Smith 1999). Without the increase in yields, it would have been necessary to bring another 3.6 billion hectares of land under cultivation, which would have almost doubled world cropland from "34 percent of the earth's surface (excluding Antarctica) to 61 percent" (Conko and Smith 1999).

The green revolution was in many ways a grain revolution—wheat, rice, and maize—but it also was a research revolution that facilitated an extraordinary and sustained general expansion in food production. Bender and Smith (1997, 18) note, "Between 1961 and 1994, the number of daily food calories per capita rose from about 1,900 to 2,600 in developing countries, while their populations nearly doubled from 2.2 billion to more than 4.3 billion."[1] Globally in the same period, "average daily per capita food supplies increased more than 20 percent." The increase in available calories per capita for the developing countries rose 50 percent from 1948–1952 to 1994–1996 (Johnson 2000, 12). A century-long trend of falling real food prices continued during the period 1950 to 1992 as international food commodity prices dropped 78 percent in constant 1990 prices (Goklany 1999, 108; Goklany 2000, 161).

D. Gale Johnson (2000, 1) sums up these achievements in agriculture: "People today have more adequate nutrition than ever before and acquire that nutrition at the lowest cost in all human history."

Green Revolution Pros and Cons

In the 1990s, while population continued to increase, the absolute number of malnourished people in the world declined by forty million, which is a continuation of a trend of the last decades. However, this was considered a "failure" by the UN World Food Program because it fell short of the reductions that were thought to be possible (AFP 1999). To the extent that it is a failure, it is a failure of institutions, not of the technology that is producing the food that could make the program a success.

So, there are arguments critical of the green revolution. The first argument is that it was a failure. Data such as that just cited above have demolished that argument, although a few still assert it, unaware of the change in the antitechnology party line. Next, we are told that the new varieties may have increased yields, but they led to a worsening of income distribution and the condition of the very poor. As early as the 1970s, studies emerged showing that the poor consumers and farmers benefited proportionately more than other groups as part of the overall improvement in human well-being (DeGregori 1985, 121; Scobie 1979; David and Otsuka 1994).

Some argued that tenant farmers in the Punjab in the north of India were displaced from their land due to high-yielding varieties of wheat. It turns out that they returned to the land as farm laborers and earned more than they did as tenant farmers. Similarly, in southern India, between 1973 and 1994, the average real income of small farmers increased by 90 percent, and that of the landless—among the poorest in the farm community—by 125 percent. Further, caloric intakes for small farmers and the landless rose 58 to 81 percent, and protein intakes rose 103 to 115 percent (World Bank 1998, 6). In addition, greater farm income meant more income and employment in the local villages that supplied the farm with goods and services now affordable to those who worked the land.

The green revolution in Indian agriculture is credited with being a major force in poverty reduction in India. From independence to 1970, the percentage of the population of India living below the absolute poverty line remained relatively steady at 55 percent. In over two decades since the introduction of the green revolution technologies, the percentage of the population of India living in absolute poverty has fallen to 35 percent (IFPRI 1999; Wood 1998, 378). Throughout most of East and Southeast Asia, the early phases of development involved an emphasis on agriculture and food production. The green revolution technologies in agriculture have played an important role in the dramatic reductions in poverty rates throughout the region, such as in Indonesia from 58 percent in 1970 to 8 percent in 1993, and in China from 270 million people in poverty in 1978 to 65 million in 1999 (IFPRI 1999). Yao (2000) estimates the reduction in rural poverty in China to be from between 596 to 790 million people (75.5 percent–100 percent) in 1978 to 57 to 114 million (6.7 percent–13.2 percent) in 1996 (447). By either measurement, the magnitude of the reduction in poverty in China is extraordinary and without precedent in human history.

In the time period 1969–1971 to 1990–1992, global population was increasing about 45 percent while the absolute number of people suffering from chronic undernourishment in developing nations decreased from 917 million or 35 percent of their population to 839 million or 21 percent (Goklany 2000, 161). From 1990–1992 to 1995–1997, the absolute number of undernourished people in the developing world continued to decline, falling to 790 million (UNFPA 1999, 3; FAO 1999). The prevalent high number should not be minimized, nor should we be complacent, but neither should the progress achieved be ignored. For we can finish the job of eliminating undernourishment by recognizing the forces that have brought us this far, thereby gaining some understanding of what still needs to be done. Unfortunately, many wish to deny the progress that has been made, offering antitechnology solutions, which would reverse these gains.

Mechanization, Income, and Erosion

Mechanization, which often came with the green revolution technologies, was said to be bad because it displaced labor. Again, it often happened that the increase in crops and output per year because of varieties that needed shorter growing seasons and higher yields per crop had labor requirements that more than offset the labor saved by mechanization. In most areas, the extra crop, thereby demanding greater labor, would have been impossible without mechanization and the green revolution technologies. In the 1970s, saving labor by mechanization was always assumed to be a bad thing, though it might also mean more leisure for the farmer. Increasingly, mechanization means that farmers may earn cash income off the farm, affording school fees for children or consumption of products in the market economy. It was conceded that the demand for agricultural labor was up, but that mechanization, by doing the more menial tasks, displaced women and children from the farm labor force. Many who oppose child labor in factories somehow deem agricultural stoop labor by children in the hot sun to be virtuous. As incomes go up and child labor is displaced, there is a very good chance that they will now be in school. And where incomes go up, other technologies are also acquired which improve the quality of life and save on child labor.

The high-yielding varieties of rice have been basically for production in irrigated fields. The upland varieties have barely changed in yields over the last three decades. This has led to the following argument:

> [The] Green revolution has given rise to greater regional inequality of income. Careful empirical studies of rice cultivation in Asia have found that laborers from the upland regions migrate to work in the paddies, smoothing out the potential regional differences (David and Otsuka 1994).

There are also arguments seeking a reduction in genetic variety, soil erosion, or groundwater contamination, which are of serious concern but cannot negate the enormous gain in human nutrition or the human catastrophe that would result from any attempt to reverse the green revolution. The diminution of genetic diversity is not a phenomenon of the green revolution per se but of agriculture. The individual farmer will select for yield or other output concerns, not for global genetic diversity. Modern no-tillage (or reduced or minimum tillage) agriculture using pesticides for weed and pest control conserves both soil and water better than its organic competitors (DeGregori 1985, 111–12).

Modern Food Supply and Safety

Regularized and improved food supply has been a consistent factor throughout history in improving human health and longevity

(DeGregori 1988). In more recent times, improved food safety has added another dimension to human health. The evidence on the safety of our modern food supply is quite contrary to the strongly held conventional wisdom. Humans have had to deal with food contamination throughout history. Human foodstuffs have been the carriers of botulism, ergot, or aflatoxins from the fungus *Aspergillus flavus,* which has caused mass illness, blindness, and large-scale death (Matossian 1989). Even uncontaminated natural foods contain substances that would be considered a threat to human life if they were used as a food additive.

In the United States, during the early twentieth century, contaminated food, milk, and water caused many foodborne infections, including typhoid fever, tuberculosis, botulism, and scarlet fever (MMWR 1999h). Once the sources and characteristics of foodborne diseases were identified, there were technologies that could help to control them before vaccines or antibiotics: "handwashing, sanitation, refrigeration, pasteurization, and pesticide application." Healthier animal care, feeding, and processing also improved food supply safety (MMWR 1999h).

Plant-produced Chemicals

The buzz word "chemicals," used by some to label and condemn food additives, could appropriately be used to describe what the plants themselves manufacture. Plants of all kinds, and most other life forms, are chemical factories, manufacturing a variety of chemicals, some of which have very active properties. Humans have extracted some of these chemicals for medicinal uses, some as poisons. Given the truism that dose makes the poison, some chemicals from plants that were once used as poisons are now used for medical purposes. In other words, chemicals from plants are not inherently good or bad; goodness or badness is determined by a variety of factors and the uses to which they are put. One of the reasons for maintaining as much biodiversity as possible is that the earth's plants produce an array of active chemicals, the vast majority of which have not been screened for possible human use. According to Katharine Milton (1993), plants are unable to run from hungry predators and so have developed a variety of defenses to avoid the loss of edible components (87). "These protections include a vast array of chemicals known as secondary compounds (such as tannins, alkaloids and terpenoids). At best these chemicals taste awful; at worst, they are lethal"(87).

Leopold and Ardrey (1972, 512) add that these plant secondary substances are "chemicals that do not participate in the basic metabolism of the plant. Among these are many chemicals that serve to repel or discourage the use of the plant by insects, microorganisms, nematodes, grazing animals, and man." These plant toxins apparently are often used to ward off fungi (Levy and Cipolini 1997). "While antifungal chemicals in nightshade were distasteful . . . there are probably protective compounds in other wild fruits that prevent rotting without spoiling taste, compounds that should be of particular interest to plant breeders" (Yoon 1997). Many plants produce "phytoestrogens," the so-called estrogen disrupters, which might have the effect of reducing the fertility of their predators, the animals that graze or otherwise consume them. "Essentially, they give their predators an oral contraceptive" (Easterbrook 1999b; Safe 1995).

Eating fresh, uncooked produce may have many taste and health benefits, but it can also be dangerous unless proper precautions are taken (Bruhn 1997; Zink 1997; Lindsay 1997; Beuchat and Ryu 1997). As epidemiologist Michael T. Osterholmm stated: "The heart-healthy diet may be kind to your cardiovascular system but it is hell on the digestive tract" (Stephenson 1997a, 97).

Rachel Carson, whose book *Silent Spring* is seen as the inspiration of the modern environmental movement, recognized the role of natural toxins—systemic pesticides—in plants. She maintained that the perception of nature as an "enchanted forest" was a "fairy tale" (Carson 1962, 32–33).[2] Even some of the weeds associated with cultivars and often eaten with them are endowed with toxins and other substances deleterious to human health. Garn notes, "Weeds growing along with cultivars often included plants with interesting alkaloids, many of them in the tomato family that includes *Datura, Belladonna,* and the like" (Garn 1994, 90).

Toxins in Plant Evolution

Through time, plants evolved for their own survival, not to serve human needs (Boyer 1982, 443; Ames 1983, 1256). "Plants in nature synthesize toxic chemicals in large amounts, apparently as a primary defense against the hordes of bacterial, fungal, and insects and other animal predators . . . *plants in the human diet are no exception.*" The evolutionary process involved in domestication changed the survival strategy for these plants, as they now depended upon the continued cultivation to survive. The toxicity of many plants was reduced compared to their wild progenitors, but in most cases was not completely bred out of plants. In fact, currently there is considerable effort to breed plants with greater disease resistance so as to reduce the need for pesticides, in some instances by increasing the production of these toxins; in other instances, not (North 1990 and Brody 1990).

Foodborne Diseases

Foodborne diseases have always been a problem for humans, even for hunting and gathering peoples, says Cohen (1989). Diseases might be caught from wild animals, including rabies or tularemia, which may have been "a significant cause of sickness and death among American Indian populations who regularly handled game and fur bearing animals." Cohen adds, "Handling wild animals or their remains can also result in infection with such other diseases as toxoplasmosis, hemorrhagic fevers, leptospirosis, brucellosis, anthrax, salmonellosis . . . people can encounter a variety of highly lethal anaerobic bacteria, including the agents of gangrene, botulism, and tetanus, if they expose themselves to the intestines of animals while butchering a kill" (33). Add in trichina worms (trichinosis) and staphylococcal infections, and it is clear that food harvested in the wild is not necessarily clean, wholesome, and natural. Drinking water in the wild can cause "beaver belly" (giardiasis) or be a source of "microorganisms derived from moose, ducks, and geese as well" (Garn 1994, 92). Water in the wild today, including snow collected from the ground and melted, can be a source of a variety of parasites, including giardia and cryptosporidium (Wellness Letter 1998, 4). If you think that we eat contaminated food today, and that we ate clean food in earlier times, think again.

Some foods require cooking or other forms of preparation to be digestible by humans, such as soaking in water, having husks or shells or other parts removed, or milling and reducing particle size to facilitate digestion. Even today, with small-scale technologies, grain milling is incomplete, leaving broken pieces of husk, causing stomach lesions and digestive problems. In the past (and in many less-developed areas today), "harvesting and milling techniques resulted in a considerable accretion of grit and little stones, as we see in the worn teeth of our colonial ancestors, Bronze-age and Iron-age Europeans, and American Indians." With storage, "weevils were often a regular part of the daily bread" (Garn 1994, 90). On the bright side, food contaminants, such as weevils, were sometimes important sources of protein in otherwise protein-deficient diets.

Health Food Fads

Health food faddists who look with horror on the minutest trace of any chemical (read manufactured chemical) nevertheless will ingest mega doses of natural products (such as herbs) that contain substances known to be toxins or untested compounds or have active products such as amino acids that have not been tested for safety at the dosage in which they are being taken (Herbert 1992). To call free-standing amino acids "natural" is a travesty, since they are not naturally present in food in significant quantity nor can they be "extracted from food without first chemically breaking food protein in the laboratory into its constituent amino acids" (Herbert 1992, 28). It is ludicrous to label a bottle of amino acids "natural," and it shows how meaningless the term has become and how useless it is for formulating public policy. The amino acid pills are a synthetic product, produced in large vats in chemical factories using microbial fermentation, with a feedstock for the microorganism that is so replete with contaminants that it has to be further processed through a carbon filter and by reverse osmosis to produce what rarely occurs in nature, a free-standing amino acid (Herbert 1992, 28). We get most all of our amino acids by ingesting complex proteins.

In 1989, L-tryptophan amino acid pills were identified as a cause of a rare muscle disorder, eosinophilia-myalgia syndrome, which eventually resulted in thirty-seven deaths and at least, 1,500 illnesses. Activists blamed it on the fact that the manufacturer used a bacteria, *Bacillus myloliquefaciens,* that had been genetically engineered, and they continue to argue this to the present, despite the fact that the overwhelming evidence is that it was a contaminant resulting from a reduction in the carbon filtration and modification of the reverse osmosis process (Mayeno and Gleich 1994). The tragic incident does not reflect well on the testing procedures (or lack of them) by the firms that bought the bulk powder for tryptophan (or other pills) and converted it into "all natural" pills.

Conveniently ignored by the activists is their long history of opposing any regulation or even monitoring of these so-called dietary supplements and their ability to pass legislation, before and after the tragedy, which specifically exempts them from precisely the oversight that might have prevented it. In fact, in 1972 the U.S. Food and Drug Administration tried to withdraw the GRAS (Generally Recognized as Safe) designation for amino acids in "dietary supplements because FDA felt that the available data did not allow" the FDA to deem them to be safe. The FDA had a number of concerns, including possible toxic or other adverse effects (USFDA 1991). Because of an error in its original GRAS designation, the FDA lost a court case that would have allowed it to have some reasonable oversight (USFDA 1991).

Vitamins

Vitamins are natural, even if manufactured, as long as they are labeled organic. They too have not been tested for safety at the mega doses that some faddists use them, though clinical studies have shown that they can be helpful when used intelligently. Obstetricians who frequently prescribe multiple vitamin pills for pregnant women and programs for vitamin A intervention have saved children's eyesight and lives in poor countries (IOM 2000b; NAS 2000c; Stephenson 2000). The faddists who would regulate agricultural chemicals out of existence oppose even simple truth-in-labeling requirements for their beloved natural products. Groups such as the International Advocates for Health Freedom, the Life Extension Foundation, and "a growing number of allies in the 'patriotic' movement" see a secret conspiracy of "international pharmaceutical empires working hand-in-hand with the UN and the FDA to restrict access to low-cost herbs, vitamins and minerals around the world" (Warchol 1998b).

Sprouts

Alfalfa sprouts have been found to be contaminated with *E. coli* O157:H7 and salmonella. Obviously, not all alfalfa and other sprouts are contaminated but the rate of contamination for raw, untreated sprouts is considerably higher than for most other food. The Emerging Infections Program of the Centers for Disease Control has estimated that twenty thousand people in North America contracted salmonella from alfalfa spouts in 1995 (JAMA 1999a; Van Beneden et al. 1999; Beuchat and Ryu 1997; Tauxe 1997a). Disease outbreaks have led to steps such as irradiating alfalfa seeds. Irradiation of alfalfa sprouts is currently being evaluated as an "adjunct seed decontamination method" (JAMA 1999a).

Raw alfalfa and clover sprouts are recognized sources of foodborne illness in the United States. . . . Raw sprouts present unique food safety problems because the warm humid conditions needed to grow spouts are also ideal for the rapid growth of harmful bacteria (CDHS 2000a,b).

The very natural nature of alfalfa sprout production creates the contamination problem and makes them a "well-suited vehicle for salmonellosis." JAMA (1999a) points out, "Seeds are often stored for months or years under cool, dry conditions in which salmonellae are stable. During the three- to five-day sprouting process, numbers of salmonellae may increase by three to four orders of magnitude, decreasing little if at all during subsequent refrigeration."

In addition, the way that alfalfa sprouts are stored, distributed, and marketed poses a danger to the health of the user:

From farm to table, many opportunities exist for contamination of alfalfa seeds or sprouts. Crops can be easily contaminated with dirty water, runoff from adjacent farms, animal fertilizers used in previous growing seasons, or droppings from rodents or ruminants (JAMA 1999a).

The natural way that alfalfa sprouts are consumed at home or even in restaurants also adds to the danger. "Alfalfa sprouts are rarely washed or cooked before consumption, and consumers are left with little protection other than chance" (JAMA 1999a). The salmonella organism is believed to be present in seed crevices between cotyledon and testa and therefore is not amenable to chemical treatment. At present there is a "fundamental problem" in that the commercial "sprouting process contains no 'kill step' that would eliminate pathogens without compromising a seed's germination potential. Consequently, because of the difficulties in producing a "pathogen-free seed," warnings have been issued about the dangers of eating alfalfa and other sprouts (JAMA 1999a; Taormina, Beuchat, and Slutsker 1999; Beuchat and Ryu 1997; Tauxe 1997a, 1998; Fett 2000).

The authors of the study conclude: "Alfalfa sprouts are a high-risk food for salmonellosis. All consumers, particularly those at greatest risk for severe disease (immunosuppressed, elderly and very young people), should consider this danger when deciding whether to eat alfalfa sprouts" (JAMA 1999a).

Proctor (1995) adds, "alfalfa sprouts contain 1.5 percent by dry weight of canavine, a highly toxic substance" (134). Not only is natural not necessarily better, food processing is an essential component of food safety. Says Garn (1994), "Some cultivars are quite toxic, unless properly prepared" (90).

Alternative Medicine

Some of the most dangerous ingested products legally sold are found in health food stores. Herbal dietary supplements (including exotic oriental herbs) and amino acids pills have been causing health problems for some of their users. A chemical additive may or may not be inactive, the herbs and amino acids are taken precisely because they are active and will alter a bodily function. Most of them are untested and unregulated. To the extent that testing is done, there is a critical difference between the approach to herbal medicine and that of pharmaceuticals. Drugs are not permitted to be used until proven to be safe and effective, while the philosophy and practice of alternative medicine has operated on the "presumption that treatments are safe and effective unless proven otherwise" (Kolata 1998).

Alternative medicine is big business. One study estimated that in 1990, "Americans poured $13.7 billion into alternative remedies, and made more visits to alternative healers than to primary-care doctors" (Freyer 1997). An editorial in the *New England Journal of Medicine* clearly defines the difficulties with alternative medicine: "What most sets alternative medicine apart . . . is that it has not been scientifically tested and its advocates largely deny the need for such testing" (Angel and Kassirer 1998).

Lacking is "the marshaling of rigorous evidence of safety and efficacy" necessary for approval of drugs and for publication in the "best peer-reviewed medical journals" (Angel and Kassirer 1998). One respected study has found that Americans are spending $27 billion a year on alternative treatments, which is close to their out-of-pocket expenditures on modern medical treatment (JAMA Reports 1998; Eisenberg et al. 1998). Some herbs and treatments were found to be beneficial, but some were not (O'Hara et al. 1998; Jacobs et al. 1998; Melchart et al. 1998; Winslow 1998; Winslow and Kroll 1998; Balluz et al. 2000). Investors have found the stocks of the firms manufacturing these dietary supplements and alternative medications to be attractive, indicating a belief that consumers will buy them, whether proved to be safe and effective or harmful. Even for some herbs that have been used in low dosage for centuries, there are no studies on the impact of the mega doses that are often taken.

The herbs and alternative medicine advocates have succeeded in blurring the line between food and drugs. Herbal extracts sold as food supplements are being added to packaged, canned, and frozen food products—called functional-foods—with labels that imply health benefits and have sufficient growth potential to have attracted some of the major food producers.

Some of the early advocates of alternative foods and medications, like Rodale Inc. and its publication, *Prevention,* recognize the danger in current patterns of use of herbs and dietary supplements. *Prevention*'s national survey on the self-care movement reveals "158 million consumers use dietary supplements for their health and spend $8.5 billion each year." It reported that "widespread use of dietary supplements may cause public health problems" (Rodale 2000; Gugliotta 2000a). There were risks mixing dietary supplements with prescription drugs and treating oneself without a doctor's supervision. The survey estimated that "11.9 million customers have experienced adverse reactions from using herbal remedies and 6.5 million have had problems of this kind when using specialty supplements" (Rodale 2000).

The problems with various herbs are legion, including lack of long-term studies and of any kind of control of dosage intake. Since many who use alternative medical treatment are also taking prescription drugs under the care of medical doctors, there may be substantial health risks, including death, in combining these with herbs and other alternative medications and stopping the use of some herbs after one had been taking them. Serious problems have emerged from ingesting some of the most popular herbs, such as *Echinacea purpurea*, St. John's Wort, or Ginkgo Biloba (Keane et al. 1999; Ernest 1999; Jobst et al. 2000; Piscitelli et al. 2000; Ruschitzka et al. 2000; Yue, Bergquist, and Gerden 2000). Probably the most serious harm can come from the interaction between herbs and prescription medicines (Lancet 2000a,b; Fugh-Berman 2000). The increasing provision of these herbs by parents to their children is evident in the statistics of the American Association of Poison Control Centers (Gugliotta 2000a,b).

One herbal remedy sometimes called ephedra, sometimes called ma huang, sometimes called pseudoephedrine, contains the chemical ephedrine, "a potent drug whether it is synthetically manufactured or produced naturally from a plant" (Burros 1993). Dr. Robin R. Caldwell says the drug is "chemically almost identical to amphetamine" (Burros

1993). Not only is ma huang sold in health food stores, it is a listed ingredient in many products sold at truck stops and elsewhere to help drivers stay awake. Pseudoephedrine is in various cold medicines that carry label warnings not to be used by anyone with high blood pressure, heart trouble, or liver problems. Ma huang is also used in diet and pep pills. Because ephedra is classified as a dietary supplement under the 1994 Dietary Supplement Health and Education Act, the FDA can restrict its sale "only if it receives well-documented reports of health problems associated with it." Brody (1999) notes, "The agency took four years, and more than 100 reports of life-threatening symptoms and 38 deaths, to act against . . . ma huang, a stimulant that can prove disastrous to people with heart problems."

Warnings have also been posted for Chinese herbal products containing aristolochia acid, which has been shown to cause cancer and kidney failure (Nortier et al. 2000; Gottlieb 2000). Nearly one hundred women with kidney damage from Chinese herbs at a Belgian slimming clinic were studied, and several were found "in dire need of kidney transplants" and many of them were developing cancers in the kidney and bladder, years after taking the drug" (Papp 2000; Greensfelder 2000).

Those opposing any regulation of dietary supplements are involved in a contradiction that doesn't seem to bother them. These herbal and other dietary supplements are taken because they have some curative effect from an active ingredient or agent. Stated simply, "if it's strong enough to work, it's strong enough to hurt" (Singleton 1999).

Technology and Life Expectancy

The presumed superiority of all things natural in food and other of life's necessities is the other side of the belief that the food supply in affluent societies is contaminated and is part of the larger, firmly held conviction that modern life is unsafe. One wonders, if contemporary life is so polluted, why are we living so long? Riley (1989) has argued that since we live longer and are older, we have more time at risk to ill health (215).

In fact, morbidity has fallen, along with mortality. We live longer, healthier lives. Among older Americans whom Riley deems to "have more time at risk to ill health," there has been a dramatic decline in chronic disability, according to the National Long Term Care Survey, as reported in a 1997 study by the National Academy of Sciences (PRB 1997, 3; ACSH 1997, 11; Crimins et al. 1997; Fogel and Costa 1997, 62). "In the last decade alone, senior citizens have experienced a 25 percent decline in the number of days of restricted activity due to illness" (Graham and Wiener 1997, 9). Improvement in the quality of life continues through the last year of life, even as we live longer (Liao et al. 2000).

In the workplace where safety has been largely ignored by environmentalists, the Centers for Disease Control and Prevention has found that "work-related deaths declined from 61 per 100,000 workers in 1913 to four per 100,000 workers in 1997" with a 50 percent decline in the risk of a fatal or disabling accident from 1970 to 1990 (Mays 1999; MMWR 1999d; Graham and Wiener 1997, 8). Says Mays (1999), "An estimated 23,000 people died from work-related injuries in 1913 compared with 5,100 in 1997, according to the National Safety Council and the U.S. Bureau of Labor Statistics. During this same period, the work force grew from 38 million to about 130 million."

Contrary to Riley, one might argue that the factors contributing to longer life are part of the same cluster of factors that reduce morbidity, including chronic illness. Fogel and Costa (1997, 56) argue cogently, and with massive data, that the epidemiology of chronic disease is not separate from that for contagious disease. Beginning in utero or infancy, inadequate nutrition can lead to a vast array (far too many to list here) of deleterious conditions that make the offspring more susceptible to contagious diseases, to chronic illness later in life, and to shorter life expectancy (56–57).

Advances in Childbearing

Giving birth to a child is dangerous to child and mother. One demographic study for the Plymouth Colony in the seventeenth century concluded that one in every thirty childbirths resulted in maternal mortality, which was about 20 percent of all female deaths. Add in the 25 percent estimated death rate for infants and children (for both sexes), and the conclusion is that close to half—about 45 percent assuming roughly equal infant and child death rates for the sexes—of all females died before completing fertility (Demos 1970, 66 and 131). This was in a society where, the author found, the "records suggest a standard of life and health that would compare favorably with that of any preindustrial society today" (Demos 1970, 65–66). Edward Shorter, in his controversial writing on the history of women's bodies, argued that virtually all women had some lifelong debilitating health problem resulting from giving birth to children (Shorter 1982; 1991). Given the high rates of maternal mortality, it is not unreasonable to assume that the debilitating injury rate was a multiple of at least five, putting the injury rate close to 100 percent.

Recent studies would seem to indicate that giving birth to a child, even under favorable conditions, limits a woman's ability to make the required "investments in somatic maintenance" for longevity and thereby shortens her life expectancy (Westendorp and Kirkwood 1998, 743; Promislow 1998). In earlier times, when nutrition levels were very much lower, and the disease experiences (both previous and current infections) of the expectant mother were much higher, it is very likely that there was organ damage to mother and child and that the mother's ability to make the "investments in somatic maintenance" for longevity were greatly diminished. If having children, even under the most favorable conditions, is harmful to the mother, then an additional benefit of lower infant mortality is that it allows a woman to achieve her desired family size at a much lower health cost to her. This clearly reinforces the thesis of Fogel and Costa (1997) on the importance of adequate nutrition and protection from disease for the long-term health of both the mother and the child.

Fogel and Costa call their theory "technophysio evolution." They say, "The theory of technophysio evolution rests on the proposition that during the last 300 years, particularly during the last century, humans have gained an unprecedented degree of control over their environment" (49). This control sets humans not only "apart from all other species, but also from all previous generations of *Homo sapiens.*" They add, "This new degree of control has enabled *Homo sapiens* to increase its average body size by over 50%, to increase its average longevity by more than 100%, and to improve greatly the robustness and capacity of vital organs" (49).

The data on the decline in maternal mortality in developed countries (and to some extent for developing countries) in the twentieth century are every bit as dramatic as the declines in infant mortality and increase in life expectancy during this same time period. The 1998 World Health Organization report confirms that the global trend toward longer and healthier lives continues, though concern is raised for the gap in life expectancy and good health between the rich and the very poor. WHO is now talking about "health expectancy" being as important an indicator of human well-being as life expectancy (WHO 1998, 58–60; Easterlin 2000). Herman (1998) says, "Life expectancy . . . in rich countries and relatively poor ones, has catapulted to marks unthinkable a century ago. In the industrial world 25 years in life expectancy have been gained in the 20th century, more than twice the gain in all previous human history."

Life Expectancy

Many of the news articles on the global increase in life expectancy discuss the problems created by this increase and the social and economic changes necessary to accommodate it. The age at which people retire, the funding of retirement and medical care systems, and the need for education and training for second careers are all issues (or problems), among many others, that will need to be addressed. In many cases they are already issues under serious discussion in economically advanced countries. The character of these systems will differ in the world in 2050, when one-fifth of the global population will be over sixty-five, from those few that existed at the beginning of the twentieth century when the first social security systems began and less than 1 percent of the world's population was over sixty-five (Herman 1998; Gelbard, Haub, and Kent 1999, 38). Though the numbers differ from country to country, the magnitude of the change for most individual countries will be roughly comparable. Technological progress does not offer the promise of a problem-free utopia, but it does offer a never ending stream of challenges with the ever-present possibility of continuing human betterment.

Water and Public Health

Humanity did not regularly drink hygienically clean water until the advent of purification processes, which included adding chemicals, in the early twentieth century. The late eighteenth and early nineteenth centuries brought developments in basic science, which allowed us to understand respiration (Joseph Priestley and Antoine Lavoisier) and statistics that laid the foundations for quantitative medicine. Lavoisier's work led to the "secularization and demystification of water by analyzing it and showing that it could be broken down into hydrogen and oxygen" (Goubert 1989, 2). In the middle of the nineteenth century, the physician John Snow identified the water from one well as the source for an outbreak of cholera, followed by Louis Pasteur's recognition of the microbial origin of many diseases, some of which were waterborne (Rockett 1994, 4–7). Previously, water could be visually clean or ritually clean, but with this new knowledge and chemical intervention, we could have hygienically clean water. In short, "water became an industrial product" (Goubert 1989, Chap. 7; Hamlin 1990, 301).

Chlorination of water in the United States began in the early part of the twentieth century and very quickly "produced dramatic reductions in morbidity and mortality associated with waterborne disease, such as typhoid, cholera, amoebic dysentery, bacterial gastroenteritis, and giardiasis." Chlorinating contaminated drinking water stopped a potential typhoid epidemic in Chicago in 1908. The introduction of drinking water disinfection in the United States is credited with reducing the incidence of cholera by 90 percent, typhoid and leptospirosis by 80 percent, and amoebic dysentery by 50 percent. A chlorinated lime solution was used as an antiseptic in hospitals in the nineteenth century. Prior to chlorination of water, diarrhea and enteritis were the third leading cause of death in the United States (Farland and Gibb 1993, 3; Burnet and White 1975, 186; Rockett 1994, 7).

Despite the fact that millions of lives have been saved by the use of chlorine for disinfection of water, there are some who would ban its further use, even though the evidence for its danger is meager (Abelson 1994a, 183). Chlorination of water is one of the triumphs of the twentieth century, but many people in the world have yet to experience its benefit. The best estimates are that 80 percent of all diseases and more than one-third of deaths in developing countries are caused by drinking or cooking with contaminated water. About twenty-five thousand children die each day from waterborne diseases (Gribble 1999, 199).

The concerns expressed about the use of chlorine are for the chlorine compounds that emerge in combination with other chemicals in the water. For any other chemical used instead of chlorine, we would have to be concerned with its toxicity *and* with the toxicity of every possible compound that it might form in the effluent from our water system. The environmentalist precautionary principle would simply rule against any substitute, particularly when set against the nearly century-long safe, effective use of chlorine.

Because of U.S. risk assessments claiming that chlorine is a carcinogen, Peru stopped chlorinating its drinking water in 1991. What followed was the "largest outbreak of cholera in recent history, which killed nearly 7,000 people and afflicted over 800,000 more" (Graham and Wiener 1997, 15; Putnam and Wiener 1997, 132, 145; Salazar-Lindo et al. 1993, 403–4).[3] Regulators, "influenced by the public's present tendency toward chemophobia," at times fail to weigh risks and benefits adequately (Malaspina 1992, 16).

An outbreak of *E. coli* O157:H7 occurred in one community in the United States where the spring water used in the public water system was not chlorinated (Rice, Clark, and Johnson 1999). In response to the outbreak of *E. coli* O157:H7, the community began chlorinating its water supply. Even with seemingly adequate chlorine levels, an indoor swimming pool complex with "extensive water spray features disseminated pathogenic bioaerosols," causing serious lung infections among its young, healthy lifeguards. Among corrective steps taken was the "provision of free residual chlorine in pool water from chlorine gas" (Rose et al. 1998, 1795, 1799).

In a study in Britain covering the years 1937 to 1986, "defective chlorination" was blamed in eight out of ten outbreaks of disease from public water supplies and in all thirteen outbreaks in private supplies (Boyce 1998b, 19). Costs of not using a technology are often not considered. The cure for not using a technology is to use more technology and use it more intelligently.

Bottled Water and Tap Water

In Houston, people are buying bottled water, which is from the same source—the municipal water supply—that they can draw safely from their own tap more conveniently, at far less cost, and without a plastic bottle that will eventually be discarded and add to environmental pollution.[4] The municipal water supply as the source for the water is clearly printed on the labels of the bottles. "In fact, 25 percent of all bottled water is tap water drawn from city water treatment plants, according to Bob Brady of the International Bottled Water Association," though the bottler may do things to the water after it is drawn (Thompson 1998). An estimate by the American Waterworks Association finds that as many as 50 percent of bottled water manufacturers in the United States obtain their water from municipal water departments. "This is the stupid thing. They buy water from the municipality for about 40 cents for 1,000 litres, they filter it two or three times, then they sell it at 2,000 times that price" (Cheney and Foss 2000).

Much of the bottled water that is not from the local municipal water supply is drawn from aquifers in rural areas. In the United States, where the legal principle of "right of capture" prevails, a major bottling company can come in, drill a few wells, and draw vast quantities of water. In Texas, there have been charges (and denials) that companies are drawing down the aquifers, causing the existing wells of local inhabitants to go dry.

The most important factor behind the extraordinarily rapid growth of the bottled water industry is the public perception that bottled water is cleaner and safer and tastes better than tap water. A study of the bacterial content of bottled water by the University of Mississippi questioned whether it is better than most municipal tap water (Cheney and Foss 2000).

Fluoride

In the United States, the average cost of tap water is one cent ($0.01) for five gallons. One very important difference is that tap water is regulated for its safety; bottled water (in Houston) is not. In addition, most bottled water lacks chemical additives such as fluoride (only about twenty brands out of more than five hundred in the United States have it), which has contributed to improved dental hygiene (Woodward 1997). Alternative, more expensive fluoride sources are, according to the American Dental Association, "not an effective or prudent public health practice." Even the International Bottled Water Association "recommends customers talk to their dentist or doctor about supplements if they are concerned about fluoride deficiency" (Woodward 1997).[5] Those who are improving their drinking water by using various purifiers and filters may be removing fluoride by using distillers or reverse-osmosis units. Speaking for the American Dental Association, Dr. Michael Easley states that he is "concerned about people who are relying on bottled water. They're not getting enough fluoride and may not realize that they are depriving their children, who will pay the price their entire lives" (Woodward 1997).

Easley points out, "At an average cost of 54 cents per person, per year, the cost of a lifetime of fluoridation for an individual is less than the cost of one small dental filling" and results in "more than $80 in avoided dental-treatment costs" per dollar spent on fluoridation (Easley 1997, 9; MMWR 1999i). Recent surveys in the United States indicate that 55 percent of children between the ages of five and seventeen have no tooth decay in their permanent teeth. Fluoridation of water is also beneficial for adults and older people. There is less tooth decay and bone loss for older people who live in communities that fluoridate their water. Some firms are marketing bottled water with fluoride added, while one health professional is recommending that parents who continue to use bottled water should consider the need for prescription fluoride as it is unlikely that fluoridated toothpastes will provide enough fluoride to maximize protection. All of which is complicated compared to just drinking tap water.

At the rate of 86.5 gallons of public water per person per day, Americans use tap water for drinking, cooking, washing dishes and clothes, bathing, carrying waste, watering lawns and gardens, and many other uses. With rare exceptions, the water emanating from public water departments is of the highest quality, having been subjected to 150 to 200 scientific quality tests, and would be the envy of our ancestors prior to the twentieth century, or of the hundreds of millions of people in the world today without access to clean water, or any reliable water source, clean or otherwise. Many Americans make a fetish of drinking bottled water, thinking they are getting a hygienically superior product.

Notes

1. For similar data, see Oerke et al. (1994), 761; Mitchell, Ingco, and Duncan (1998); WHO (1998), 130; and Goklany (2000), 161.

2. Quoted in Rubin (1994), 49. Carson also recognized the need for some use of pesticides to control insects (Carson 1962, 12-13, cited in Rubin 1994, 33).

3. One study of this epidemic finds 20,000 deaths and over a million cases of cholera (Gribble 1999, 200).

4. On the quality of drinking water in the United States, see Symons (1992), Symons (1995), and Lalumandier and Ayers (2000).

5. On the safety of fluoridation, see Cook-Mozaffari, Doll, and Kinlen (1993).

CHAPTER 9

Technology: Control for, or over, Humans?

Technology is a human endeavor that involves artistic insight, science, technique, experimentation, and risk. The risks inherent in any experimentation are compounded by an apparent propensity to reach beyond our grasp. To Henry Adams (1963), it was building the cathedral Mont-Saint-Michel so "high on the summit of this granite rock" that the walls collapsed several times (1). In our own time, it includes our breaking the bounds of earth's gravity and reaching for the heavens. To Sir Alfred Pugsley, "a profession that never has accidents is unlikely to be serving its country efficiently" (Petroski 1985, 221). Among his contemporary writers, Faulkner (1956) rated Thomas Wolfe number one because he failed. "He had failed best because he had tried hardest." Ernest Hemingway he rated last "not on the value of product at all but simply because Hemingway having taught himself a pattern, a method which he would use without splashing around to try to experiment." Hemingway did not "attempt to reach the unattainable dream." In his acceptance speech for the Nobel Prize for literature in 1949, Faulkner said,

> I decline to accept the end of man. . . . I believe that man will not merely endure: he will prevail. He is immortal, not because he alone among creatures has an inexhaustible voice, but because he has a soul, a spirit capable of compassion and sacrifice and endurance.

In art, as in technology, or as in any human endeavor, never to have failed is life's greatest failure. For it means not having pushed oneself, or civilization, to the limits of achievement of that time. One only knows these limits by trying to go beyond them. In the creation of art or technology, we are creating ourselves. Unlike other species that derive their nature from their biological being, we are an unfinished product, never to be perfect or complete, always in the making. Human achievements derive further meaning as standards to be built upon and records to be broken.

The question about technology being good or bad has less meaning than asking the question about how we can understand and make better use of it. We evolved along with our technology and are unthinkable without it. Technology is not external to us. To say that technology is alien is meaningless. Our very biological being emerged as a result of an interactive process with the technology that we were creating. As early hominids used their technology to leave the tropics, the ability to inhabit and survive in new environments depended upon the creation of new technologies.

Just as the technological process that allowed us to become *Homo sapiens* was irreversible, so was and is the technology that has allowed us to grow in number from a species of about one to five million a short thirty thousand years ago, to six billion people today. "Invention is the mother of necessity," as Veblen's dictum claimed (Veblen 1922, 314). Returning to an earlier technology is simply not an option. Intelligent criticism of technology today must be framed so as to define a technological alternative, or pathway to a technological alternative, that sustains human life at its present, and likely its future, level.

The success of the pessimists' propaganda can be illustrated by the fact that at a time in which virtually all indices of human well-being in the United States show highly positive change and when the vast majority of Americans consider their own personal condition and prospects to be good or very good, polls also show that "national optimism has never been lower, most people thinking that the country is in decline" (Easterbrook 1999a, 19). This has been called the "I'm okay, they're not okay" syndrome—also called the "optimism gap" (Whitman 1998).[1] Such unwarranted pessimism is widely believed across the political spectrum, with Left and Right having their own reasons for positing a civilization allegedly in decline. The doomsday mentality has been reinforced by the media; bad news sells newspapers and magazines and gets good TV ratings. Madigan (1999) says, "Presenting a steaming cornucopia of dysfunction ranging from presidential misbehavior to local crime stories, police reality shows and talk shows where the unspeakable becomes common, the media picture of life can be grim" (Madigan 1999).

Except for two events, the Associated Press top ten stories for 1998 were "gloom, doom, scary sex and even scarier weather" (Madigan 1999). Similarly, a list of journalism's one hundred "greatest hits" of the century had only two items even remotely considered science and technology: John Hersey's *Hiroshima* and Rachel Carson's *Silent Spring*. The assembly line and the mass production of goods such as the automobile are not included, but Ralph Nader's *Unsafe at Any Speed* is (Barringer 1999; DeGregori 1999). Was science and technology's main twentieth-century accomplishment to find new, more efficient and more horrible ways for us to kill each other? There was no airplane, no space stories, no polio vaccine, no insulin, no antibiotics, no eradication of smallpox, nor any of the century's innumerable other triumphs to qualify as news. Radio, television, and photography made the one hundred list only as stories about war or domestic conflict, and not about inventions or improvements (DeGregori 1999).

Science and technology always need criticism, both by those developing a technology and by disinterested scientists and technologists external to a particular project. Criticism of a new technology is most effective when it is based on solid scientific judgment and when it is offered by those who would welcome the benefits derived from a technology but have reasonable doubts about it. The interaction between the developers of a technology and its reasoned critics creates safer technology and accelerates the pace of technological change. As a development economist, I know that even worse than not introducing a technology is introducing the wrong technology or introducing the right technology at the wrong time or in the wrong way. Each failure is immediately costly to human well-being, and makes it more difficult at a future time to promote the right technology that serves a vital human need.

Free and open inquiry is the lifeblood of science and technology. The open or public nature of inquiry works only insofar as there are rules of the game whereby ideas and practices are tested and either falsified or, in not being falsified, are at least provisionally accepted as being true. If not fully accepted, an idea that continues to successfully resist falsification is deemed to be the best that we have available and what we choose to act upon. Whatever its weaknesses, the scientific method is still the best way we have to resolve differences, allowing us to get on with the processes of life. Brown (1999) describes the scientific method as a "process still being invented" and considers it the greatest invention of the last two millennia. Whatever shortcomings one finds in scientific knowledge, it must never be forgotten that the "starting point for science was not complete knowability . . . but complete ignorance." Science has reduced "the uncertainty of a world that was at first totally inexplicable" (Budiansky 1995, 207). Those opposed to modern science believe that the twentieth century movement in science away from absolutism somehow justifies their postmodernist claim that no body of knowledge is "privileged." Budiansky responds, "Humans have never been able to predict the natural world with certainty, nor have scientists assumed that they could . . . theoretical limits imposed by quantum mechanics and chaos theory changes nothing" (207).

Following from scientific inquiry is verifiable knowledge. Knowledge has been defined as "incomparably greatest among human resources. . . . It is greatest because it is the mother of other resources" (Mitchell 1941, 1; DeGregori 1987a, 1243; 1996). The scientific method may not be the "mother" of all other knowledge, but it has certainly become the midwife. Beyond being a method for exploration and acquisition of knowledge, the scientific method has become the most effective operational tool for sorting out testable truths from the massive influx of data from the natural and human-made environment.

As with all human endeavors, the processes of science and technology are not without problems. Critics have pointed to many elements of bias, which should be taken seriously and corrected. But the postmodernists have not even remotely come up with anything better. Their criticisms serve as a basis for improvement—continuing to "invent" it—not for abandoning the scientific method. "Scientific 'truth' is indeed not fixed and absolute 'out there' in the world waiting to be discovered, but *a special kind of relationship between the knower and the known*" (Fox 1996, 337).

A Warranted Assertion

The very skepticism implicit in scientific inquiry is one of its greatest strengths in a democracy. Ours is not a quest for absolute values but a process of what John Dewey (1939) called valuations or judgments that we must make because life does not give us the luxury of not making choices. "All scientific claims, of course, are shot through with cultural expectations, but science itself is a self-correcting process that over time weeds out the culturally contingent for the universal" (Lewis 1996, 216). Like science, democracy is a self-correcting system.

Dewey and Charles Sanders Peirce before him were proposing what Peirce (1992) called "thought in action" (129). Thought in action produces "a rule of action" or "habit" in the "exercise of volition" that temporarily "appeases the irritation of doubt" (129). The essential key, to Peirce and Dewey (and what we are advocating), is that because "belief is a rule of action, the application of which involves further doubt and further thought," action becomes simultaneously a "stopping-place" and a new "starting-place for thought" (129). To Peirce, if our actions become increasingly untenable, then the "irritation of doubt" will grow and give an even greater impetus to innovative thinking. Though there is and should be an element of doubt in all our actions, we can still act with very reasonable and very strong confidence that if our actions are in terms of this process, we fallible human beings are doing the very best our current state of knowledge allows us to do. This is the supreme mark of intelligent behavior, and little more can be asked of us except to continue to strive to do better.

In Chapter 1, John Maddox argued that even more exciting than finding answers to questions now "crying for our attention" is the excitement that will come from finding answers to questions that "we do not yet know enough to ask" (1). As long as the process of thought and action continues, then so will the "irritation of doubt" and the new ideas and theories that will arise from them and from our excitement at their discovery. If in fact today's truths are someday looked back upon as quaint and possibly even wrong, it will not invalidate what we are doing today. Today's knowledge may be overturned tomorrow (or it may not be), but it is today's ongoing technological and scientific inquiry that creates the pathway to a future stream of new understandings. Tomorrow's truths will be as provisional as today's truths. This will not invalidate what we have done in our time. Despite totalitarian regimes and misuse of technology, the record shows we have overwhelmingly used today's technology and science as we have understood them to increase human life and the quality of life at a rate and magnitude never previously achieved in human history.

In our age of seeming uncertainty or at least of continuous change that frequently challenges established ideas, those who do not wish to respond to change, or simply fear it, must turn to dogmas that promise absolute certainty and protection against change. The point is not that there is more uncertainty today, but that there is a greater understanding and recognition of uncertainty. Science and technology provide greater areas of predictability in our everyday lives than ever before but have made us more aware of areas in which great uncertainty remains and have also given us greater possibilities for change, therefore greater freedom. For many, this greater freedom is the real threat.

Not only will we never have perfect predictability and perfect order, it is not clear that we want it. Though few would admit to wanting such a condition when so starkly stated, it is nevertheless an unstated and unrecognized assumption for advocates of the precautionary principle who demand that we guarantee with certainty the safety of a technology. A large measure of order is necessary in our knowledge, our personal lives, and our society, for us to function intelligently. It is equally true, though, that a certain amount of disorder is necessary in every system, physical, social, and intellectual, if novelty is to arise. No matter how far the frontiers of knowledge may be pushed, no matter how much predictability we achieve, there is, in all of life (including doing nothing) an element of risk and uncertainty. One dimension of maturity in the modern world is to recognize both what we are capable of achieving and what life's risks and uncertainties are. As humans, we have no choice but to live with risk and uncertainty.

Dewey's "warranted assertion" is provisional. A warranted assertion is sufficient for us to act in terms of it, but not to stifle the arguments against it except by reasoned discourse. To Dewey (1938), knowledge is a "continuing process in every field in which it is engaged" (8). Postmodernists and fundamentalist critics alike have often tried to interpret the lack of finality in human inquiry as either an argument for postmodernist relativism or for fundamentalist absolutism. It is neither. It is simply the best way that we have thus far found to carry out the process of problem solving and bettering the human condition. In the long term, it is the only way we can exist and coexist in a world of diversity.

Defense of Technology

A difficulty for the public understanding and defense of technology from its critics is that technology is simultaneously ubiquitous and invisible. It is ubiquitous in that, from the moment we awake until the time we go to sleep, technologies from recent decades or, at most, the last century, are an inescapable part of everything we do. If anything goes wrong, we can blame the technology. Some of the most important technologies are largely invisible. They are, like freedom of speech, not noticed until someone tries to take them from us. Most of us were immunized in a distant past at an age that has passed from memory. This almost invisible shield of technologies shelters and protects us so efficiently and effectively that it escapes our notice.

One of the innumerable examples of technology changing our lives in ways that are not fully appreciated can be found in a study by two demographers. Their research raises and answers the question as to "how many Americans are alive because of twentieth-century technology?" (White and Preston 1996). They found that if the mortality rates of 1900 had prevailed throughout the twentieth century, half of all Americans alive would not be here today. Half of that half would have been born and already died. This means that many older Americans would not be alive. The other half of that half or "another quarter of the population would not have been born because their parents, grandparents, or earlier ancestors would have died before giving birth to them or their progenitors" (White and Preston 1996, 428). This factor would disproportionately impact on younger Americans who would never have been born. Further, since 67 percent of the effect on population was from a decline in mortality for the zero to fourteen age group, many of today's young people would have been born but died in infancy or childhood (White and Preston 1996, 426). Overall, who is in what "category is impossible to say," as the 50 percent figure is an average that "cannot be assigned to individuals. Because this 50 percent estimate shows little variation by age or sex, any individual could sensibly estimate whether he or she is alive due to survival improvements by merely flipping a coin" (White and Preston 1996, 428).

Mortality Rates

Globally, 25 percent of today's population would not be alive without the decline in mortality rates in the last fifty years. Declining fertility rates have about balanced the declining death rates so that today's population is slightly less than it would be if both fertility and mortality rates had remained unchanged for the last half century. More births but people living much shorter lives was transformed into fewer births

but those born living longer lives. If current trends continue, population will be just under 9 billion in 2050, while the forward projection
of the 1950 fertility/mortality rates gives a population of 15 billion
(Heuveline 1999, 690). Another dramatic twentieth-century change in
survival rates is that for many developed countries, "the chance of survival from age 0 to age 60 is now greater than was the chance of survival from age 0 to 1 in 1900" (Castles 1998, 834).

As we have noted, all in the United States have benefited from the
twentieth-century advances in technology. That does not mean that all
today share equally in its benefits, merely that all groups have experienced improvement. The inequalities of today are of long standing.
This means that we have two ongoing obligations in both seeking to
advance science and technology to continue to improve the overall
health and well-being of the community and work equally as hard for
fuller participation and access to these improvements for everyone.
The numbers cited above would become even more extraordinary if all
had shared more fully in the change. We don't help the disadvantaged
by opposing the technology that has helped us all. We must act, both
in advancing technology and in enhancing its availability to as much of
humanity as possible.

As White and Preston (1996) tell us, "mortality reduction throughout the world has been more rapid in the twentieth century than in any
previous period" (415). In the United States in 1900, one infant in ten
died in the first year and another one in ten died before age five for a
total of one in five infant and child deaths before age five (CDC
2000a). Equally as spectacular as the decrease in infant and child mortality rates since 1900 has been the one hundredfold decline in maternal mortality in the twentieth century (Moore and Simon 1999, 9;
MMWR 1999e; CDC 1999d). The last figures available before the new
millennium showed that life expectancy continues to increase, and infant mortality continues to fall in the United States (Guyer et al.
1999).[2] Yet the idea prevails among many that modern technologies are
killing us, despite the incredible twentieth-century increases in life expectancy in the United States and around the world. The one question
that I repeatedly ask in this book and elsewhere is, if technology is
killing us thus, why are we living so long?

The global rate of change in mortality and morbidity coming later
and moving more rapidly is in many ways more astounding than that
for the United States and other developed countries. During the course
of the twentieth century, average life expectancy in the world has more

than doubled from thirty years to over sixty-six years (McFalls 1998, 13). Over the last four to five decades, life expectancy has increased about twenty years (USAID 1998, 1; Goklany 2000, 171). From 1955 to 1995, life expectancy increased ten years, from sixty-seven to seventy-seven years in developed countries, while it went up twenty-four years from forty to sixty-four years in developing countries, which translates into a 60 percent increase (WRI 1998, 8). Every one-year increase in life expectancy "would appear to be accompanied by slightly more than a one year improvement in Disability-Adjusted Life expectancy (DALE)" (Murray and Lopez 1996a, 238). "The proportion of expected life span lived with disability declines as life expectancy rises" from a high of 15 percent to around 8 percent (Murray and Lopez 1996a, 238; WHO 2000d,e). If the current evidence is confirmed, as life expectancy increases, the proportion of life with disability decreases, and the absolute number of years with disability decreases.

In 1950, 287 out of every 1,000 children in developing countries died before their fifth birthday. In 1995, the under-five mortality rate for developing countries had fallen to 90 (and to 88 in 1996), a decline of about two-thirds (WRI 1998, 8; UNICEF 1998, 97). For roughly the same period, 1955 to 1997, the absolute decline was from 21 million to 10 million (or 11.7 million in 1996 according to UNICEF) in the number of children dying before reaching age five (USAID 1998, 1; UNICEF 1998, 97). Since world population more than doubled during this time period, between 42 and 45 million children under five would have died in 1997 if 1955 death rates prevailed, or better than four times the actual number.

Since 1950 the total number of people dying in the world has hovered around 50 million, though population has much more than doubled. There were several years between 1950 and 1980 when the number of people dying was actually fewer than in 1950. Currently, estimates of the number of people dying in the world each year are around 52 million. If 1950 death rates prevailed today, well over 50 million more people would die each year, with over 30 million being children, and with total deaths being considerably over 100 million each year. An important component in the decline in death rates has been the decrease in deaths from infectious diseases (CDC 1999c). Currently, about 6 percent of all deaths in developed countries and 34 percent of all deaths worldwide (or 17.25 million) are from infectious diseases (Murray and Lopez 1996a, 176, 181; WRI 1998, 11; USAID

1998, 1). More recent figures indicate that communicable diseases may account for 23 percent of the global disease burden (Reddy 1999). The latter figures are consistent with the International Red Cross/Red Crescent estimate of 13 million deaths from infectious disease (out of just over 50 million deaths a year), which they find to be rising and mostly preventable (IFRCRCS 2000a). In the past century infectious diseases ceased to be the cause of the preponderance of human mortality. We have won a lot of victories in this battle against infectious diseases, but the war is far from over. Resistance to treatment is increasing for some diseases and new diseases are emerging, giving us no cause for complacency (Smith et al. 1999; Waldvogel 1999).

Gains in Reducing Mortality

Despite these very serious problems, gains continue to be made in reducing morbidity and mortality from infectious diseases. Though communicable diseases account for only 34 percent of all deaths, they still are responsible for 54 percent of years of life lost (YLL). The best estimates show that lifestyle choices are a significant factor in premature deaths from diseases (McGinnis and Foege 1993). Earlier work by the famed epidemiologists Richard Doll and Richard Peto (1981) indicate that lifestyle choices account for a significant portion of cancer deaths, with at least 30 percent of all cancer deaths being the result of the use of tobacco. Kinlen (2000) argues that "apart from the elderly, in whom death certification of cancer has improved, age-specific mortality rates from cancer as a whole are now lower than they were in 1911 (the earliest available), and but for lung cancer, would be much lower" (122). There is cause to celebrate the victories we have won, as long as we are determined to continue to disseminate the benefits that we have achieved and remain active promoting research to defend the gains we have made and to continue the struggle for longer healthier lives for all of humanity.

Just as our economic indicators probably greatly understate the real change in our standard of living, the numbers on the extraordinary changes in mortality and life expectancy almost certainly understate the magnitude of the change for both developed and developing countries. In the United States in 1900, many infants born in rural areas who died within a few days of birth were simply never recorded as births, except maybe in the family Bible. In the last fifty years, vital statistics for developing countries have been standardized and greatly improved.

In many cultures and countries, a child was not considered to have been born unless he or she had lived for a culturally determined number of days and therefore was not counted unless he or she survived that period.

The good news about life expectancy seems to be getting better. The rate of increase in the second half of the twentieth century was faster than that for the first half and, according to new projections, may be faster in the future than are shown by the estimates that have been used for the current calculations of life expectancy (Tuljapurkar, Li, and Boe 2000; Horiuchi 2000). Living longer presents a variety of problems of aging, particularly financing retirement, unless people are willing to retire later or save more while working. Unfortunately, the data that I have just cited also show that AIDS is taking its toll in many poor countries, particularly in Africa (where it has surpassed malaria as the leading cause of death), with both life expectancies and disability adjusted life expectancy falling in some countries. Much needs to be done to reverse this trend.

Under the heading of "children are living healthier and safer lives than ever before," Kimberly Thompson (2000) describes the following difference between a "day in the life of typical child 100 years ago" and one today. For example, "today's American child":

- can expect to live approximately 30 years longer on average
- is six times more likely to finish four years of high school
- will never know polio or spend time in an iron lung
- won't get tuberculosis from milk, rickets from vitamin D deficiency, scurvy from vitamin C deficiency, or cretinism from lack of iodine in the diet. (1)

Thompson goes on to list the diseases that children today are protected against: "diphtheria, pertussis, measles, mumps, rubella, hemophilus influenza B, hepatitis B, and chicken pox." And they will have been "screened for conditions like phenylketonuria, congenital hypothyroidism and lead poisoning." There are "antibiotics to treat an infectious disease" (1). Among the infectious diseases being treated is pneumonia, for which the absolute number of children dying fell 97 percent, from 24,637 in 1939 to 800 in 1996, while the child mortality rate from other causes was falling 82 percent (Dowell et al. 2000, 1399). In 1900, "30.4 percent of all deaths occurred among children aged less than 5 years; in 1997, that percentage was only 1.4 percent.

In 1900, the three leading causes of death were pneumonia, tuberculosis (TB), and diarrhea and enteritis, which (together with diphtheria) caused one third of all deaths. Of these deaths, 40 percent were among children aged less than 5 years" (MMWR 1999e).

Raising the Standards

What technology has done in this instance, as it does in many others including other factors in child mortality, is that it raises our standard as to what is acceptable. Infant mortality rates of one hundred per thousand births were acceptable though deplorable in 1900 because that was close to what our technology allowed us. Infant mortality rates in the teens per thousand births as experienced by some minority groups in the United States are unacceptable to many of us because they are often twice that of the dominant white majority, even though these rates for minorities are but a small fraction of those that prevailed for everyone in 1900. That technology raises our standards or values as to what is acceptable, is not to deny the importance for social change of the moral values and decency of those who tirelessly work for the betterment of their fellow human beings. It is a word of warning, however, to those well-meaning individuals who mistakenly join forces with the Luddites, who promise much and deliver little since opposition to technological change undermines the most effective and essential means for positive social change.

Facing Facts

If the critics of modern technology are meaningfully to engage the noncritics, they must at least accept the overwhelming data on improvements in human life that have been taking place and continue to do so. In other words, chemicals aren't killing us; they are part of the larger process of science and technology that has been saving and improving our lives. A particular chemical or class of chemicals may be killing us in the sense that the harm it causes may outweigh the benefits it is supposed to provide. Before we restrict or ban a technology, we also need to know who gains and who loses from using it or not using it, and whether there are particular uses for some people where the gains are greater than the losses. Slogans or simplistic dichotomies do not contribute to problem solving. Neither does fear!

Within the framework of the overall benefits of modern science and technology, there is still a very broad arena in which intelligent human beings, using verifiable knowledge, reason, and the scientific method of inquiry, can differ on the formulation of public policy on technological issues. Such a public discourse can lead to greater advances in human well-being. Accepting the overwhelming data on the benefits conferred upon us by modern technology and science provides a perverse twist to clearly demonstrated cases of a chemical in use causing harm or a given number of deaths being iatrogenic—the result of faulty medical practice. If we have achieved such extraordinary gains in spite of some errant chemicals not being caught by our regulatory process, or in spite of medical practices in need of improvement, the potential gains from the rest of technology and science and their more careful utilization are even greater, not less. It means that we need not less technology but more intelligent use of it.

Intelligent Use

Those of us who defend technology have to recognize that the intelligent use of a problem-solving instrument is as much a part of the technology as is the hardware itself. Intelligent use involves a range of social and ethical issues of consideration of the rights of others. A technology with enormous potential for improving one's life, such as a cellular phone, can be used by some in public in ways that can legitimately be considered an annoyance to others. New technologies bring change that requires a continuous renegotiation of the rules by which we live together. In many cases, the benefit or problem-solving characteristic of a technology was not what the inventor or discoverer intended and was, in fact, what was invented by its subsequent use. As I have argued, tools become technology when used and in a very real sense, a technology is created and recreated every time it is intelligently used (DeGregori 1977). Every intelligent, humane use of a technology with even a modicum of novelty is part of a larger process of creating new technology for of all of humanity.

Balancing Needs and Rights

Technology may not have separated us from the "natural world," but our technology and the affluence it has created has sheltered us from

its harsher realities. Some of us only know animals as pets, not as a basis for livelihood or simply survival. We have never seen the family
farm blown away in a dust storm and with it all the hopes and dreams
of years of hard work. This was a reality for some of our grandparents
and remains a reality for many poor farmers throughout the world. Nor
have we had to watch helplessly as insects or disease devoured our
crop and with it our family's survival. Some humanists can oppose the
construction of dams for irrigation and power or the use of pesticides
for crop protection. For them, electricity is flipping a switch, water
flows with the turn of a faucet, food is a trip to the supermarket, health
food store, or fast food outlet, and one's family's future is a series of
paychecks with a pension at the end of the line.

I have seen alienated intellectuals routinely oppose dam construction in poor countries that may desperately need reliable sources of water and power. Building a dam may mean flooding lands that are used
for grazing by nomadic peoples. The issue is complex, trying to balance competing claims for equity and development. Intervention by international groups ideologically opposed to dam construction unduly
complicates the issues further. These groups use their influence in the
capitals of developed countries to block funding by donor or lender organizations, as has been done with increasing frequency in recent
years. Donor agencies lacking a major domestic constituency in developed countries often yield to small pressure groups who are adept at
using the media to promote their antidevelopment cause. Various
rivers' websites allow dam opponents to learn why they are opposed to
construction on a river that they couldn't find on a map. I wonder what
moral right these groups have to deny water to poor farmers or electricity to power a pump, a radio, or a few light bulbs?

Guthrie's Vision

Construction of the public works in developed countries, including
dams, highways, and rights of way for power and telephone lines, have
involved the same complex questions of whose rights are restricted,
who gains, and what is fair compensation for those who lose. It also
should involve sensitivity to special claims, such as land that may be
sacred to a people. Not too long ago, radical populists, like Woody
Guthrie, were protechnology. If Guthrie were alive today and sang
"Roll On Columbia," he would probably be booed by the ideological

descendants of those who earlier cheered him. Guthrie had a vision of the rivers of the American northwest "tamed" by technology—dams. In "Roll On Columbia," the Grand Coulee Dam was "the mightiest thing ever built by a man" and would power "the great factories and water our land." Unlike some of today's intellectuals, Guthrie knew the hard scrabble of Oklahoma farmers in the Great Depression. The power from a dam "turning our darkness to dawn" was important to someone who grew up without electricity.

Guthrie's vision of irrigation and cheap power running factories, filling "shiploads of plenty," and curing the economic problems of the depression, may seem naive in retrospect. But at least he had a vision. It was a radical vision of "mighty men" who "labored by day and night, matching their strength against the river's wild flight through rapids and falls." On the face of it, the strength of people is insignificant, compared to that of a river, but working together with technology, we humans can turn the river's power to our use. As Guthrie claimed, "they won the hard fight." However naive this lost vision may be, it still has much to offer the world today. If we are going to work together to address the many problems of poverty and the environment, we need a vision like that of "Roll On Columbia" so we do work together, not against one another.

A great populist senator from Texas, Ralph Yarborough, campaigned on the slogan of putting the jam on a shelf where more people can reach it. The vision claimed by many antitechnology partisans cannot be realized unless we allow the jam to be made and put on the shelf.

Humanitarian Aid

Antitechnology criticism reaches its true nadir when humanitarian aid for people in need is attacked because of the technology used to produce it. In India, following a super-cyclone, a team from Vandana Shiva's research foundation gathered samples of donated grain while involved in "relief work" and had them tested in the United States to see if they were genetically modified. Claiming that they were genetically modified, Diverse Women for Diversity then demanded that the government of India "immediately withdraw the corn-soya blend from Orissa." It appeared that they preferred starvation for the cyclone victims to a presumed but unproven contamination from GM food (RFSTE 2000; Devraj 2000; Lean 2000; Jayarsman 2000c).

Science Education

There is clearly a problem in science education. The public may not know less science than earlier generations; but there is so much more to know to be an informed citizen. The issue is further complicated when ideological interest groups, such as those against genetically modified food crops, carry on an expensive, highly organized campaign of disinformation. Some of the beliefs that the zealots feed upon would be hilariously funny if they did not have such adverse implications for public policy. The following example is illustrative.

In 1999, polls found that a majority of the British and European public either believed (or didn't know) that "ordinary tomatoes do not contain genes while genetically modified ones do" because they had been added through biotechnology, and that genes from genetically modified plants could become part of our genetic heritage if we ate genetically modified tomatoes (STN 1999; Hoban 1999; 2000). A news story began by quoting a consumer saying that one thing she could not abide was genes in her tomatoes. The ability of a citizenry to act intelligently in their own best interest in the formulation of public policy in a democracy is threatened by a lack of understanding of science and technology and makes a portion of the public susceptible to the blandishments of ideologues.

The Realistic View of Life

We can best create the world as we want it to be only if we have a realistic view of the world as it is. We argue against the romantic view of other cultures or earlier times not because we are mean-spirited or chauvinistic. Rather, we are trying to focus attention on the possibilities inherent in the world that we already inhabit. Our exaltation of the present is not meant to be a denigration of the past. However high the infant and child mortalities may have been or how short the life expectancies, those who have gone before us had the courage to continue and find meaning and value to the life process. As Camus (1955) has argued, suicide is the most important and serious question in life because we as intelligent sentient beings have the power to terminate our own at any time. That we don't was an affirmation of life in all its joys and sorrows, triumphs and tragedies and an affirmation of the will to continue. "It is bad to stop, hard to be satisfied with a single way of seeing, to go on without contradiction . . . the point is to live" (65). Referring to the Greek myth of Sisyphus forever rolling a rock toward the top of a mountain, only to have it roll back before reaching the summit, Camus adds that the "struggle itself towards the heights is enough to fill a man's heart" (123).

Those who went before us created the life that we now can lead, as we must and likely will create a better world for those who follow us. Our respect for other cultures and peoples in earlier times and different places does not mean that we have an option to return to the ways that have characterized the human condition. Having tasted the rich fruit of longer healthier lives and the myriad other benefits of the present, we would find only misery where others had once found acceptance, if not occasional joy. There are many ways forward from which to choose, but the one choice we don't have is to return to a nonexistent idyllic past.

The world today is a wonderful, exciting place filled with a seemingly infinite supply of experiences to satisfy each of us through several lifetimes. It is also a place replete with problems to be solved, the solution of which can also be among life's most fructifying experiences. Romantic obfuscation of reality gets in the way of both intelligent problem solving and the realization of the fullest potential of the life process, both individually and collectively. Those who speak of the "reenchantment of the world" miss the point that, as Postrel argues, we already "live in an enchanted world, a world suffused with intelligence, a world of our own making." It is not a world of limits but one of "plenitude" where there "lies an adventurous future" (Postrel 1998, 217–18).

Forecasting the Future

"Ready or not, computers are coming to the people" (Brand 1972, 50). Thus, Stewart Brand began a prescient 1972 article in *Rolling Stone* when few had access to computers and the personal computer did not yet exist. He spoke of the time "when computers become available to everyone." Those using computers would "be all the more empowered as individuals and as cooperators." This "would enhance things . . . like the richness and rigor of spontaneous creation and of human interaction . . . of sentient interaction." His most perceptive observation was that "until computers come to the people, we will have no idea of their most natural functions" (58). Brand was ahead of his time and ahead of ours as we have just begun to exploit information technology for human betterment.

Even the most insightful and accurate observer of change, such as Brand, must recognize that there are limits to forecasting the future. This is consistent with my underlying theme of emergent evolution. Emergent evolution argues that an emergent system, be it a biological organism or a new technology, will have properties and potentialities that cannot be fully understood prior to the emergence of the system. In philosophy, the theory has definite problems, as it implies there are effects that do not have a cause. This problem does not exist for knowledge about technological evolution and the social change it engenders, since it does not matter whether the emergent properties were indeterminate in any causal sense, or whether it is effectively impossible for humans to gather all the information and to formulate the theory governing the new system before it arises, even if it is otherwise fully determined. Stated differently, the evolution of a technology such as computers, the personal computer and the interconnections that led to email, the worldwide web, and global access to information, also created new system properties with which most of us were unfamiliar and therefore were almost completely unable to understand until they happened. To many, this is a cause for concern sufficient to severely restrict the evolutionary processes of change. We recognize that change can be adverse. Our position is for individual and system openness to new possibilities, with a recognition that we have to understand new systems and conditions as they emerge in order realize their best potential.

Bountiful Future

The technologies at our disposal and the cumulative and accelerating nature of technological change offer a promise of an ever more bountiful future. Ours is an enthusiasm, almost a wild enthusiasm, tempered with reasonable concern. The model that we have offered is one of a problem-solving process in which each new solution creates its own problems. That is both the promise and challenge of technology. Technology does not offer us utopia, but many people don't want utopia. They prefer the continued challenge of problem solving, believing that not only can we rise to new challenges, but also in the process we can and will continually rise above them. Eternal vigilance may be the price of liberty; it is also the price of technological progress and of life itself.

Basic optimism does not mean we simply let the future "happen" since we know it will be better. Quite the contrary, as I have indicated repeatedly, the presumed certainties of earlier times were an illusion, as death and destruction visited individuals and communities far more frequently and capriciously than they now do. Today, we have more knowledge of the forces that drive the future than ever before, and it is our responsibility to shape them for human betterment. Most of our serious actions involve the future in some sense. Research in science and technology, creating infrastructures or opportunities, has its future embedded in them, and how we carry it out will have great impact on the possibilities for future improvements.

Our having vastly greater knowledge, allowing us to shape the future, does not mean that we can totally determine it. The greater knowledge we have makes us more keenly aware of the uncertainties than in earlier times, when such matters were thought to be under the aegis of a larger but incomprehensible force or forces, whether benevolent or malevolent. The way we view the world today will have an enormous impact on the shape of the world tomorrow. People with an optimistic view of life have one great fear, that the pessimistic worldview can become a self-fulfilling prophecy. Policies to restrain technological, resource-creating change and have us "live with limits" will create these limits and will eventually exhaust possibilities within them (DeGregori 1985).

The principles that the pessimists proclaim, such as entropy, however they may misunderstand them, define and create a truly dismal future for humanity. The point is not whether resources and the stuff of life's potential are infinite or finite (the mathematical concepts do not apply here), but whether they are open-ended. In other words, is there the capability, as far as we can meaningfully peer down the corridors of time, to create an ever richer and more meaningful future? Our answer is, unequivocally, yes! It matters greatly, then, whether we see ourselves as merely drawing from nature's diminishing bounty or whether we see ourselves as active creators, or at least cocreators, in a dynamic partnership in which we humans are the active force for change.

The fear of an imminent exhaustion of critical resources seems to have abated, though there are still some holdout believers. Almost without exception, the real prices of commodities, including mineral resources, are at historic lows and, according to an early 1999 study by the World Bank, resource prices are likely to remain low, even with economic recovery in Asia. The study indicates that the technological change in the efficiency of resource discovery and extraction has been increasing resource availability more rapidly than the demand for them (Solman 1999). We can shape a more bountiful future only if we believe that we can do it. This is not a romantic statement about will and determination but is based upon the record of technological change and human advances that we have presented. The past is a very important ingredient in understanding the future and provides a solid basis of hope.

Problem Solving That Creates Problems

Earlier, I discussed the benefits of chlorination in providing clean water. Clean water was part of a package of public health measures that contributed to dramatic reductions in mortality and morbidity. It fits the model of problem solving, creating new problems. Because of advances in public health, polio became a disease whose incidence and severity increased with improved health and living standards (Burnet 1975, 91). Cohen (1992) notes, "The organism that causes the disease is normally a fairly common inhabitant of children, causing mild diarrhea. . . . But as hygiene improves, individuals on the average are infected progressively later in life" (63).

Being infected at a late age likely results in paralysis making it "primarily a disease of wealthier individuals . . . in societies that maximize the hygienic isolation of their young" (Cohen 1992, 63). Because of improvements in the standard of living, such as cleanliness and toilet training in children, and housing and sanitation, the number of children who escape infection in the "safe" period of infancy increases (Burnet 1975, 95). Having experienced the virus as mild diarrhea or subclinical infection provides later protection against having it in a more virulent form. For a forty-year period in the United States, from around 1917 to the development of the Salk vaccine in the 1950s, followed by the Sabin vaccine, polio was the dread disease of middle-class Americans (Blume and Geesink 2000).

The relationship of improved hygiene to polio was not fully understood until about the time the first vaccines were discovered. Even had it been known, there was no prior hygienic or unhygienic utopia to which we could return. The only real option was to discover or create a vaccine, improve upon it (as a relatively continuous process), and then spread it as widely as possible. This is what was done. Not only has polio been eliminated in the United States, but it is no longer extant in the Western Hemisphere and is largely confined to South Asia and parts of Sub-Saharan Africa. A global eradication campaign has been in process, aiming for complete eradication by the year 2000 or as soon thereafter as possible (Marwick 2000b; MMWR 2000a; CDC 2000d). In this case, not only did humanity rise to the challenge of the problems created by the new technologies of sanitation, it rose above them with one of the great triumphs of all time.

If one follows the antitechnology line on the internet, there is a clearly stated or implied belief that the main need for new technology is to solve the problems created by modern technology itself. In other words, return to an earlier technology (such as organic agriculture) where a safe, bountiful future awaits you. As the case of polio indicates, there is a very small measure of truth in the first part of the claim in that new technologies create new problems. However, these technologies, such as cleansing our water supply, are warranted in their use, if the problems that they create are less than those they solve.

Is continued technological optimism warranted? In my judgment the evidence presented above clearly substantiates it. One can write quite accurately of all the failures of modern science and technology, but the fact remains that in aggregate we are much better off, and there is little if any evidence that the very few who are worse off would be in a better situation in the absence of technological change. It is revealing how critics of technological optimism use the same limited number of examples of technological failure to illustrate their concern. In agriculture, critics of the green revolution raise the specter of the Irish potato blight in the 1840s or the corn (maize) blight in the United States in 1970. The Irish famine of 1848 was truly a catastrophe, but it was not a unique event in human history or something that could have been avoided prior to modern times. Opponents of technological change have gone from catastrophic projections of famine and proclamations of the failure of the green revolution to the current claim that GM food crops are not needed because the world produces enough to feed everyone. The corn blight in the United States ended up reducing yields by less than 25 percent, leaving a vastly greater output than was ever achieved before the hybrid varieties. Nothing in traditional agriculture even remotely approximates the success of modern agriculture, even during what is seen as its failures.

Critics of technological optimism like to quote two utopian technological prophecies of the 1950s and 1960s: that nuclear power would be so cheap that we would not need to meter it, and that we would have eliminated infectious diseases (Amyes 2000). Clearly the utopian prophets were wrong, but at least they understood the direction of change correctly. The gains made during the twentieth century in the fight against infectious diseases would have seemed utopian, had they been predicted in 1900. If one looks at forecasts based on data, not prophecies, one finds that those dealing with actual data have not been overly optimistic. Quite the contrary, the positive changes that we describe in this book have occurred at a pace that has been faster than the most optimistic among us would have dared to predict. In other words, one might find, at the beginning of the twentieth century, serious forecasts of life expectancy increases or infant mortality decreases not even remotely close to those that have actually occurred. Go back to 1950 and find any forecasts that would have coupled forecasts of the rapid increase in population with the even more rapid increase in food supply or the declines in mortality and in fertility. Or return to 1960 and find any forecasts of rates of economic growth for Asia that are anywhere near what has been realized. There are no impediments in finding pessimistic forecasts; it would be interesting to learn of data-based forecasts that closely approximated these outcomes.

Concern is often expressed for the "unintended consequences" of technological change, with most, but not all, assuming that what we don't expect to happen is necessarily bad. Tenner (1996), who dwells on such matters, does have material on the unexpected good things that occasionally happened. His book *Why Things Bite Back: Technology and the Revenge of Unintended Consequences* and the publicity it aroused reinforce the belief that unforeseen outcomes are likely to be bad or worse. The evidence we have marshaled argues quite the contrary. Put in proper perspective, the idea of unintended consequences is useful if it helps us to achieve an understanding of the types of actions that can engender adverse results, so that we can better avoid them or minimize their impact. It also helps to understand why potential benefits have often been ignored for some time, so that we have more power to exploit beneficial outcomes.

The Speed of Change

Few can be more certified as a technological optimist than I, having imbibed a philosophy of technological change as an undergraduate and maintained a lifelong commitment to it. Yet I have had to admit that the pace of change in the world over the last decades has exceeded even my wildest expectations (DeGregori 1998, 853). Veblen recognized that "imbecile institutions" could bring technological change and all its benefits to halt. We have made progress reforming such "imbecile institutions," but much work remains to be done. Given the progress that has occurred, as we begin the twenty-first century, it does not take blind faith to believe, as I argued here, that technological progress will continue.

Technological change is open-ended and there is no reason to expect that it should at any time come to a halt. The cumulative nature of the process means that each new generation will be faced with unprecedented opportunities. Equally, though, they will be faced with unprecedented challenges. Emergent evolution and the evidence we offer argues that life will continue to get better, but we cannot predict the new emergent possibilities for human betterment. Ours is not a book of futurism or prophecy except to argue that technological change is a process of human betterment.

Instrumental Rationality

The twentieth-century modernist perspective has been characterized by what Inglehart (1997) calls "instrumental rationality." It was and remains a belief, warranted by experience, that problems exist to be solved. Maybe not all problems can be solved, maybe not right away, but enough problems are being solved that our lives are regularly improved. Operative human intelligence continues to expand the boundaries of knowledge. The vision has been valid for the arts and humanities, where it is believed that education will help to create more tolerant, open societies.

The vision of instrumental rationality in many respects fueled the very finest aspects and the highest aspirations of the twentieth century. When polio emerged as a crippler of children, research scientists set to work to find a cure. It took a while, but the cause went forward successfully. Study after study has determined that scientists place great emphasis on primacy in discovery, but still, many working on a cure for polio or other problems must have known that, though they were building the foundations upon which later breakthroughs would be constructed, their work would be largely unheralded. Simply stated, no one works hard to build a foundation unless they have the modernist faith that a worthwhile construct will arise upon it. In this regard, the vision was altruistic, not selfish, and it was socially conscious in the best sense of that term.

Inglehart (1997) contrasts modernism with postmodernism by showing that the postmodernists have a "diminishing faith in science, technology, and rationality" (79). If anything, this is a substantial understatement of the postmodernist position, as so many of them are militantly against science, technology, and rationality, which they see as being "dehumanizing" and a product of the "hegemonic" culture of the West. Inglehart notes that postmodernism has gripped those who are secure and who take their survival for granted. They live in societies that "have attained unprecedentedly high life expectancies and unprecedentedly high levels of subjective well-being" (Inglehart 1997, 78). In other words, they are condemning the forces that have benefited them, and they are restricting access to these same benefits by those who have not been so privileged.

Inglehart uses the terms *materialist* and *postmaterialist* interchangeably with the terms *modernist* and *postmodernist*. I need here only to reiterate the ideas of John Dewey and Hu Shih, quoted at length in Chapter 1. Dewey (1929) spoke of the condition of mankind, the "powers which environ him and determine his destiny," and the active human intelligence constructing the means to "turn the powers of nature to account" and use them to construct "a fortress out of the very conditions and forces which threaten him" (3). Inglehart is, in effect, resurrecting the old dichotomy between the materialist West as contrasted with the spiritualist East. It is false, as I have shown, because the East has taken to twentieth-century science and technology and excelled at it. There is an added ironic twist in that so many postmodernists presume to be speaking on behalf of Asians and other non-Western peoples. More to the point, Hu Shih (1928) says,

> That civilization which makes the fullest possible use of human ingenuity and intelligence in search of truth in order to control nature and transform matter for the service of mankind, to liberate the human spirit from ignorance, superstition and slavery to the forces of nature, and to reform social and political institutions for the benefit of the greatest number—such a civilization is highly idealistic and spiritual (27).

Overcoming the Forces

There is no better example of overcoming the forces that environ humans, as Dewey described it, than the American space program. The forces that held us on the earth were understood and used to allow us to break free and, literally, to soar to the moon. Ansel Adams, an artistic genius, gave us the justly famous photograph, "Moonrise over Hernandez, New Mexico." But only the space program could give us "Earthrise," the picture of astronauts on the moon with earth rising behind them. That one photograph poignantly captured the sense of one earth and the unity of life upon it, with the immensity of the dark space and the universe surrounding it. It will undoubtedly be one of the most memorable images of the century as it reflects one of humanity's noblest achievements and the realization of some of its highest aspirations. From *Voyager*, launched in 1979, to *Galileo* now exploring the planets of our solar system, we have been able to get a closer look at other planets and their moons and have found, among other things, the extremely active and incredibly beautiful Io, one of the moons of Jupiter. *Galileo* has transmitted some fantastic pictures from the side of Io not visible from earth (Crystal 2000). The driving force for this achievement was science, technology, engineering, organization, but it was also ideas, vision, hard work, dedication, and commitment—and it was profoundly spiritual.

Modernism has not been without its faults. Its very successes may have engendered some hubris and led to excesses and overconfidence in everything from the environment to curing disease. Like all human endeavors, it has to be continually examined and modified. But until something better comes along, there is no basis for abandoning it. Virginia Postrel (1998) is correct when she so eloquently asserts that the "future will be as grand and as particular as we are." She envisions a future of increasing possibilities and diversity. For there is no "single bridge from here to there, for neither here nor there is a single point." And above all, "there is no abyss to cross" (218). There will always be many bumps on the road ahead, detours that go nowhere and barriers that force us back a bit. This is reality, but it does not eliminate the fact that, despite the many impediments, the movement will mostly be forward as it has been throughout the past century. We will have economic advance, but we also will have to forgo some of the improvement, to deal with problems we create.

Ecumenicism

Orlando Patterson (1994a) speaks of ecumenical America, and an ecumenical, global culture that goes beyond merely being "cosmopol-

itan" and the "simple embrace of many cultures." It is a "genuinely ecumenical one: it draws from everywhere" (24). One can interpret "ecumenical" many different ways, all of them good. I like to interpret it
in terms of the American philosopher William James, when he preferred to think that we lived in a "pluriverse" not a universe. In the
United States, we can be 100 percent American and still have an ethnic heritage that adds to our being without subtracting from our Amer-
icanism. Having more diverse kinds of heritages means that we have
more to offer to our fellow citizens. In general, we are born into an ethnic heritage and either born into, or legally accepted into, a national
political system or heritage. The new global culture is one we can all
participate in, if we are able, without requiring being born or initiated
into it and without diminishing what else we are. Whatever you call it,
global or ecumenical culture, it opens the most incredible range of possibilities for each of us, in our being and in our becoming, for creating
a rich and fulfilling existence.

If, as Turner (1993) defines it, cultures are ways that specific social
groups have used the means at hand to "make themselves," then true
multiculturalism can be "envisioned as the empowerment of the basic
capacity for self-creation . . . for all members and groups of society"
(Turner 1993, 426–27). To build on Turner, a critical dimension of
multiculturalism is the process of facilitating access to technologies
that provide all members of a community the fullest "capacity for self-
creation." For us not only to coexist in multicultural societies in a multicultural world but also to draw strength from it, we have to develop
what, in another but structurally similar context, Peter Galison (1997,
795) calls a "'protocol language' to serve as a common referent" relating the multiplicity of values and ideas underlying our differences.
Technology is an integral part of any protocol.

Defining Boundaries

Unfortunately, there is no formula or a priori framework by which
we can precisely define the boundaries between differences that must
be tolerated within a political system and between those diverse practices that violate more basic human rights. The issues involved are
complex. Political slogans about the battle between absolutism and relativism obscure, rather than clarify, the issue. Nor are assertions about
the validity of cultural relativism helpful, because, taken to an extreme
(which few people actually do), relativism deprives us of the right to

criticize any practice whatsoever that is sanctioned by a culture. Recognizing that we do not have all the answers to these questions is the first step in devising the means by which we can work together to negotiate the boundaries between the permitted, tolerated, and encouraged differences and those that are unacceptable. This will be an ongoing process as time and circumstances work to alter the boundaries. Though we can formulate guidelines or create a framework within which the decision-making process operates, we will always be faced in all areas of the human endeavor with the necessity of weighing evidence and making choices.

Galison (1997) has another metaphor applied to physics and engineering but bursting with meaning for other areas of inquiry: "Physicists and engineers harbored a profound mistrust of disorder. They searched for reliability in crystals rather than in disordered materials, and strength in pure substances rather than in laminated ones" (843).

Galison notes that the "classical vision had it backwards." Crystals failed "catastrophically" because of their order, and it was the "amorphous semiconductors with their *disordered* atoms, that gave the consistent responses needed for the modern era of electronics." In time, structural engineers learned the same lesson. The strongest materials "were not pure—they were laminated" (843). Galison's use of Charles Sanders Peirce's image of the cable is even more evocative and fraught with meaning for understanding the multicultural world we inhabit. As we all know reflexively, a chain is no stronger than its weakest link. To Peirce, our reasoning "should not form a chain" but be like a cable "whose fibres may be ever so slender, provided that they are sufficiently numerous and intimately connected" (843–44). Galison continues, "With its intertwined strands, the cable gains its strength not by having a single, golden thread that winds its way through the whole. No one strand defines the whole" (844).

It is the "interweaving of many limited strands, no one of which held all the weight" that gives steel cables their strength. "It is the *disorder* of the scientific community . . . the *dis*unification of science—the intercalculation of *different* patterns of argument—that is responsible for its strength and coherence" (844). Substitute science and scientific community with society, and Galison's statement goes to the heart of the argument that I am advancing in favor of technology and modern life. Many romantics who bemoan the alleged homogenization of modern life nevertheless seek the presumed purity of an early time or of some "primitive" peoples. Their quest for societal purity is as dysfunctional as is the quest for purity in materials that Galison describes above. Like the many strands of a cable, diversity in a community is its source of strength.

Pro-technology, Pro-people

To be pro-technology is to be pro-people. Being pro-technology does not involve a denial of the real problems that we face, be they environmental or simply continued poverty amid the growing affluence. But it does mean having a balanced perspective. If we don't dwell on the negative, it does not mean that we don't recognize its existence. We do not accept the presumed conflict between technology and the arts but recognize that they are integrally intertwined, as each sustains the other. Technology gives us ever-increasing means to address the problems of the future. Technology is at the core of the human potential for a life of richness and fulfillment that can someday be available to all. And that is yet another of the glories of technology. Mine is a thesis that might be called technological populism. It can allow us to create mass culture, which I defend, and opens the possibility for mass participation in all its benefits. A human good for all humanity can be humanity's highest achievement. T.S. Eliot (1952, 145) had this to say:

> We shall not cease from exploration
> And the end of all our exploring
> Will be to arrive where we started
> And know the place for the first time.
> Through the unknown, remembered gate
> When the last of earth left to discover
> Is that which was the beginning.

Notes

1. Quoted in Easterbrook (1999a), 19.
2. For an overall summary by the Centers for Disease Control and Prevention of the twentieth century's achievements in public health, see MMWR (1999a,b,c,d,e,f,g,h,i,j,k,l); CDC (1999a,b,c,d; 2000a,b,c); and NIAID (2000).

References

Contact http://www.uh.edu/~trdegreg/ for additional references.

AAP. 2000. Vaccines Are Safe and Effective: Evidence Confirms No Link between Autism and Vaccines. Washington, D.C.: American Academy of Pediatrics, Press release, 6 April.

Abbey, Edward. 1968. *Desert Solitaire: A Season in the Wilderness.* New York: Ballantine Books.

Abelson, Philip H. 1990. Testing for Carcinogens with Rodents. *Science* 249(4975): 1357.

———. 1994a. Chemicals: Perceptions Versus Facts. *Science* 264(5156): 183.

———. 1994b. Adequate Supplies of Fruits and Vegetables. *Science* 266(5189): 1303.

Abelson, Philip H., and Pamela J. Hines. 1999. The Plant Revolution. *Science* Special Issue, Plant Biotechnology: Food and Feed, 285(5426): 367–68.

Acacio, Brian D., Tamar Gottfried, Robert Israel, and Rebecca Z. Sokol. 2000. Evaluation of a Large Cohort of Men Presenting for a Screening Semen Analysis. *Fertility and Sterility* 73(3): 595–97.

ACS. 2000. U.S. Cancer Occurrence and Death Rates Continue to Decline. ACS News Today (The American Cancer Society), 14 May.

ACSH. 1997. *Of Mice and Mandates: Animal Experiments, Human Cancer Risk, and Regulatory Policy.* New York: American Council on Science and Health, July.

Adams, Henry. 1918. The Dynamo and the Virgin. In *The Educaton of Henry Adams: An Autobiography* by Henry Adams. Boston: Houghton Mifflin.

———. 1963, orig. 1905. *Mont-Saint-Michel and Chartres.* Boston Houghton Mifflin.

Adams, Jonathan S., and Thomas O. McShane. 1996. *The Myth of Wild Africa: Conservation without Illusion.* Berkeley: University of California Press.

AFP. 1999. 790 Million Malnourished People in the World, U.N. Says. Agence France-Presse, Nando Media online, 14 October.

Agaisse, H., M. Gominet, O. A. Oekstad, A.-B. Kolstoe, and D. Lereclus. 1999. PlcR Is a Pleictropic Regulator of Extracellular Virulence Factor

Gene Expression in *Bacillus thuringiensis*. *Molecular Microbiology* 32(5): 1043–53.

AgWeb. 2000. UK Ag Minister No Health Justification for Ban on U.S. Beef. Pro Farmer Editors *AgWeb.com* online, 13 June.

Alsop, Joseph. 1982. *The Rare Art Traditions: The History of Art Collecting and Its Linked Phenomena Wherever These Have Appeared.* New York: Harper & Row & Princeton University Press, Bollingen Series 35, Mellon Lecture Series in the Fine Arts.

Altman, Lawrence K. 1990. Study of Three Mile Island Accident Finds Negligible Increases in Cancer. *New York Times*, 1 September.

———. 1998. Good News from the Front in the War Against Cancer. *New York Times*, 26 May.

Aly, Gotz, Peter Chroust, and Christian Pross. 1994. *Cleansing the Fatherland: Nazi Medicine and Racial Hygiene.* Trans. Belinda Cooper; foreword by Michael H. Kater. Baltimore: Johns Hopkins University Press.

Ames, Bruce, Margie Profet, and Lois Swirsky Gold. 1990a. Dietary Pesticides (99. 9% all natural). *Proceedings of the National Academy of Sciences USA* 87: 7777–81.

———. 1990b. Nature's Chemicals and Synthetic Chemicals: Comparative Toxicology. *Proceedings of the National Academy of Sciences USA* 87: 7782–86.

Ames, Bruce, Renae Magaw, and Lois Swirsky Gold. 1990. Ranking Possible Carcinogenic Hazards. *Science* 236 (4799): 271–80, 17 April 1987, reprinted in *Readings in Risk*, edited by Theodore S. Glickman and Michael Gough. Washington, D.C.: Resources for the Future.

Ames, Bruce N. 1983. Dietary Carcinogens and Anticarcinogens: Oxygen Radicals and Degenerative Diseases. *Science* 221(4617): 1256–64.

———. 1992. Pollution, Pesticides, and Cancer. *Journal of AOAC International* 75: 1–5.

Ames, Bruce N., and Lois Swirsky Gold. 1990. Too Many Rodent Carcinogens: Mitogenesis Increases Mutagenesis. *Science* 249(4972): 970–71.

———. 1991. Natural Plant Pesticides Pose Greater Risks Than Synthetic Ones. *Chemical Engineering News* 69: 48–49.

———. 1993. Another Perspective . . . Nature's Way. *Consumer's Research* 76(8): 20.

———. 1997. Pollution, Pesticides, and Cancer Misconceptions. In *What Risk?* edited by Roger Bate. Boston: Butterworth-Heinemann.

———. 1998. *Misconcenceptions about Environmental Pollution, Pesticides, and the Causes of Cancer.* Dallas: National Center for Policy Analysis, NCPA Policy Report No. 214, March.

Amyes, S. G. B. 2000. The Rise in Bacterial Resistance: Is Partly Because There Have Been No New Classes of Antibiotics Since the 1960s? *BMJ (British Medical Journal)* 320(7229): 199–200.

Anderson, C. 1998. Business, Health Advocates Stress Importance of Food Irradiation. Associated Press, Nando.net online, 8 April.

Anderson, K. E., J. D. Potter, and T. M. Mack. 1996. Pancreatic Cancer. In *Cancer Epidemiology and Prevention*, 2d ed., edited by David F. Schottenfeld and Joseph F. Fraumeni, Jr. New York: Oxford University Press.

Andrews, Wyatt. 2000. Green Terror. *CBS Evening News: Eye on America*, 10 and 11 April.

Angel, Marcia, and Jerome P. Kassirer. 1998. Alternative Medicine: The Risks of Untested and Unregulated Remedies. *New England Journal of Medicine* 339(12).

AoAI. 1905. Art of the American Indian. *Brush and Pencil* 15: 84–93.

Apple. Rima D. 1989. Patenting University Research: Harry Steenbock and the Wisconsin Alumni Research Foundation. *ISIS* 80(303).

———. 1996. *Vitamania: Vitamins in American Culture*. New Brunswick: Rutgers University Press.

Apse, M. P., G. S. Aharon, W. A. Snedden, and E. Blumwald. 1999. Salt Tolerance Conferred by Overexpression of a Vacuolar Na+/H+ Antiport in Arabidopsis. *Science* 285(5431): 1256–58.

APSnet. 1999. American Phytopathological Society, APSnet Feature, Press release, 15 October.

Aristotle. 1958. *The Politics of Aristotle*. Translated by Ernest Barker. New York: Oxford University Press.

Arluke, Arnold, and Boria Sax. 1992. Understanding Nazi Animal Protection and the Holocaust. *Anthrozoos* 5(1): 6–31.

———. 1995. The Nazi Treatment of Animals and People. In *Reinventing Biology: Respect for Life and the Creation of Knowledge*, edited by Lynda Birke and Ruth Hubbard. Bloomington: Indiana University Press.

Arnold, Steven F., Diane M. Klotz, Bridgette M. Collins, Peter M. Vornier, Louis J. Guilette, Jr., and John A. McLachlan. 1996. Synergistic Activation of Estrogen Receptor with Combinations of Environmental Chemicals. *Science* 272(5267): 1489–92.

Arthur, Charles. 2000. GM Feed "Will Not Give Rise to Superbugs." *The Independent (London)*, 23 March.

ASCB. 2000. Statement in Support of Research on GMOs, The American Society for Cell Biology, 26 May.

Ashraf, Haroon. 2000. African Leaders Discuss Ways to "Roll Back Malaria." *Lancet* 355(9214): 1528.

ASM. 2000a. War on Bacteria Could Leave Drug-resistant Strains Unchecked, American Society for Microbiology Press release, 18 July.

———. 2000b. Statement of the American Society for Microbiology on Genetically Modified Organisms, American Society for Microbiology, July.

ATS. 2000. New Findings on Asthma and Allergies at American Thoracic Society (ATS2000) Meeting, Toronto, Canada: American Thoracic Society's 96th International Conference 5–10 May.

Attaran, Amir, and Rajendra Maharaj. 2000. DDT for Malaria Control Should Not Be Banned. *BMJ* 321(7273): 1403–5.

Attaran, Amir, Donald R. Roberts, Chris F. Curtis, and Wenceslaus L. Kilama. 2000. Balancing Risks on the Backs of the Poor. *Nature Medicine* 6(7): 729–31.

Avery, Dennis T. 1993. Cattle Are the Green Revolution. *Earthwatch* 12(5): 26–28.

———. 1998a. Draft EPA Food Safety Leaflet Misleads Consumers, Pushes Organic Foods, Though Manure-Borne Toxins Can Endanger Kids. *Global Food Quarterly* (25), Summer.

———. 1998b. The Hidden Dangers in Organic Food. *American Outlook Magazine* 1(3): 19–22.

———. 1999. The Fallacy of Organic Utopia. In *Fearing Food: Risk, Health and Environment*, edited by Julian Morris and Roger Bate. Boston: Butterworth-Heinemann.

Baeyer, Hans Christian Von. 2000. The Lotus Effect. *Sciences* 40(1): 12–15.

Bagga, Dilprit, Karl H. Anders, He-Jing Wang, Erika Roberts, and John A. Glaspy. 2000. Brief Communication: Organochlorine Pesticide Content of Breast Adipose Tissue from Women with Breast Cancer and Control Subjects. *Journal of the National Cancer Institute* 92(9).

Bailey, Ronald. 2000. Bio-Invaders: Are We under Attack by "Non-native" Species? Should We Care? Reason online August/September.

Balluz, L. S., S. M. Kieszak, R. M. Philen, and J. Mulinare. 2000. Vitamin and Mineral Supplement Use in the United States: Results from the Third National Health and Nutrition Examination Survey. *Archives of Family Medicine* 9(3): 258–62.

Banthia, Jayanthia, and Tim Dyson. 1999. Smallpox in Nineteenth-Century India. *Population and Development Review* 25(4): 649–80.

Barnes, Douglas F., Robert van der Plas, and Willem Floor. 1997. Tackling the Rural Energy Problem in Developing Countries. *Finance & Development* 34(2): 11–15.

Barringer, Felicity. 1999. Journalism's Greatest Hits: Two Lists of a Century's Top Stories. *New York Times,* 1 March.

Bate, Roger. 1997. The Virtues of Playing with Fire. *Financial Times (London) Weekend*, 20–21 September.

———. 2000. Europe, Factory for Food Scares. *Wall Street Journal*, 23 March.

Bateson, P. P. G. 1999. Genetically Modified Potatoes. *Lancet* 354(9187).

Bazzaz, F. A. 1987. Experimental Studies on the Evolution of Niche in Successional Plant Populations. In *Colonization, Succession and Stability,* edited by A. J. Gray, M. J. Crawley, and P. J. Edwards. Oxford: Blackwell Scientific.

BBC. 1998a. Study Rejects Safety Fears. BBC World Service online, 30 April.

———. 1998b. Childhood Cancers May be "Linked to Chemical Exposure at Birth." BBC World Service online 27 October.

———. 2000. Organic Food "Proven" Healthier. BBC World Service online, 3 January.

Beachy, Roger. 2000. Testimony: U.S. Senate Subcommittee on Inter-national Economic Policy, Export and Trade Promotion. 12 July.

Bedford, Helen, and David Elliman. 2000. Concerns about Immunization. *BMJ (British Medical Journal)* 320(7229): 240–43.

Belasco, Warren. 1989. *Appetite for Change: How the Counterculture Took on the Food Industry.* New York: Pantheon Books.

Belluck, Pam, and Christopher Drew. 1998. Tracing Bout of Illness to Small Lettuce Farm. *New York Times*, 5 January.

Bender, William, and Margaret Smith. 1997. Population, Food, and Nutrition. *Population Bulletin* 51(4).

Berg, Alan. 1973. *The Nutrition Factor.* Washington, D.C.: The Brookings Institute.

Beringer, John. E. 1999a. Keeping Watch over Genetically Modified Crops and Foods. *Lancet* online 353(9153).

———. 1999b. Cautionary Tale on Safety of GM Crops. *Nature* 399(6735): 405.

Berman, Morris. 1981. *The Reenchantment of the World.* Ithaca: Cornell University Press.

Bernal, J. D. 1971. *Science in History.* Cambridge: MIT Press.

Beuchat, Larry R., and Jee-Hoon Ryu. 1997. Produce Handling and Processing Practices. *Emerging Infectious Diseases Journal* 3(4).

Biehl, Janet, and Peter Staudenmaier. 2000. *Ecofascism: Lessons from the German Experience.* San Francisco: AK Press.

Billington, David P. 1983. *The Tower and the Bridge: The New Art of Structural Engineering.* New York: Basic Books.

Billington, David P., and Robert Mark. 1984. The Cathedral and the Bridge: Structure and Symbol. *Technology and Culture* 25(1): 37–52.

Birnbaum, Larry. 1995. CD Reviews: Swingers & Singers. *Downbeat* 62(2): 46–47.

Bloom, Floyd. 2000. Rice, Races, and Riches. *Science* 288(5468): 973.

Blume, Stuart, and Ingrid Geesink. 2000. A Brief History of Polio Vaccines. *Science* 288(5471): 1593–1594.

BMJ. 1969. DDT in the Environment. *British Medical Journal* 4: 446.

———. 1997. DDT (dicophane) and Postmenopausal Breast Cancer in Europe: Case-control Study. *BMJ (British Medical Journal)* 315(12 July): 81–85.

————. 2000a. Research Does Not Reflect Global Disease Burden. *BMJ (British Medical Journal)* 320(7244): 1228.

————. 2000b. News: Mosquitoes Prefer Pregnant Women. *BMJ (British Medical Journal)* 320(7249).

Boime, Albert. 1985. William Blake's Graphic Imagery and the Industrial Revolution. *Arts Magazine* 59(10): 107–19.

————. 1987. *Art in an Age of Revolution 1750–1800*. Chicago: The University of Chicago Press.

Bond, Michael. 2000. Dr. Truth. *New Scientist* 164(2218): 74–77.

Book, Esther Wachs. 1997. In the Chinese Grain: Collectors Are Drawn to Ming Furniture. *Far Eastern Economic Review* 160(29): 53–54.

Borg, Kevin. 1999. The Strange Career of DDT: Experts, Federal Capacity and Environmentalism in World War II. *Technology and Culture* 40(4): 770–96.

Borlaug, Norman E., and Christopher R. Dowswell. 1988. World Revolution in Agriculture. *1988 Britannica Book of the Year,* 5–14. Chicago: Encyclopedia Britannica, Inc.

Boseley, Sarah. 2000. Malaria Summit Opens to Grim Statistics: Malaria's Toll on the African Economy Has Been Far Worse than Expected, Harvard Economist Jeffrey Sachs Tells the First African Malaria Summit. *Electronic Mail & Guardian*, 25 April.

Botkin, Daniel B. 1990. *Discordant Harmonies: A New Ecology for the Twenty-first Century*. New York: Oxford University Press.

Bouchie, A. J. 2000. Bacillus Identity Crisis. *Nature Biotechnology* 18(8): 813.

Bowden, Rebecca. 1999. Royal Society Rejects Latest Claims on GM Potatoes. *The Royal Society (London)*, Monday, 11 October.

Boyce, Nell. 1998a. A Necessary Evil: Banning DDT Might Seem Like a Good Idea in the North, but There Is No More Effective Weapon for Fighting Malaria in the Tropics. *New Scientist* 157(2120): 18–19.

————. 1998b. The Demon Drink: Disinfecting Water Has Saved Countless Lives Across the World. But as Recent Studies Show, It Carries Its Own Health Risk. *New Scientist* 157(2143).

Boyer, J. S. 1982. Plant Productivity and Environment. *Science* 218(4571).

Bradbury, Jane. 2000. Genetically Modified Mosquitoes Could Lead to Better Malaria Control. *Lancet* 355(9222): 2227.

Bramwell, Anna. 1989. *Ecology in the 20th Century: A History*. New Haven: Yale University Press.

Brand, Stewart. 1972. Spacewar: Fanatic Life and Symbolic Death among Computer Bums. *Rolling Stone* 123: 50–58.

Breyer, Stephen. 1993. *Breaking the Vicious Circle: Toward Effective Risk Regulation*. Cambridge: Harvard University Press.

Brody, Jane E. 1990. As Its Virtues Emerge, the Potato Goes Global. *New York Times*, 9 October.

———. 1994. Personal Health: Fears of Food Irradiation Often in the Way of Rational Consideration of Its Benefits. *New York Times*, 12 October.

———. 1998. Personal Health: Health Scares That Weren't So Scary. *New York Times*, 18 August.

———. 1999. Natural? Safe? Drug Free? It Might Be None of the Above and Americans Gamble on Herbs as Medicine. *New York Times*, 9 February.

Brough, David. 1999. Scientists in Italy Study GM Rice. *Reuters News Service* online, 14 December.

———. 2000. Biotech Can Cut World Hunger. *Reuters News Service* online, 15 September.

Brown, Peter. 1999. Science in Modern Life. *Sciences* 39(1): 4.

Browning, Robert. 1951. Andrea Del Sarto. In *The Selected Poetry of Robert Browning*, edited by Kenneth L. Knickerbocker. New York: The Modern Library.

Bruhn, Christine M. 1997. Consumer Concerns: Motivating to Action. *Emerging Infectious Diseases Journal* 3(4).

Brundtland, Gro Harlem. 1998. *Launch of Roll Back Malaria*. New York: United Nations.

Budiansky, Stephen. 1995. Chaos in Eden. *New Scientist* 148(1999): 33–36.

Burkhardt, Frederick Henry, ed. 1952. *The Cleavage in Our Culture: Essays in Scientific Humanism in Honor of Max Otto*. Boston: Beacon Press.

Burnet, F. M., and David White. 1975. *Natural History of Infectious Disease.* 4th ed. Cambridge: Cambridge University Press.

Burnham, G. 1998. Onchocerciasis. *Lancet* 351(9112):1341–50.

Burros, Marian. 1993. With Some Widely Sold Dietary Supplements, What You Don't Know Can Hurt You. *New York Times,* 14 July.

Burros, Marian. 1998. F. D. A. Issues Warning on a 60's Health Food. *New York Times,* 30 September.

Butler, Declan. 1999. Biotech Industry Seeks "Honest Brokers." *Nature* 398(6726): 360.

Byerlee, Derek, and Carl K. Eicher, eds. 1997. *Africa's Emerging Maize Revolution*. Boulder: Lynne Rienner Publishers.

Cairns, John, Jr. 1999. Absence of Certainty Is Not Synonymous with Absence of Risk. *Environmental Health Perspectives* online 107(2).

Calabrese, E. J., ed. 1994. *Biological Effects of Low Level Exposures: Dose-response Relationships*. Ann Arbor: Lewis.

Calamari, D., L. Yameogo, J.-M. Hougard, and C. Leveque. 1998. Environmental Assessment of Larvicide Use in Onchocerciasis Control Programme. *Parasitology Today* 14(12): 485–89.

Campbell, Bernard. 1982. *Humankind Emerging*. Boston: Little, Brown & Co.

Campion, Edward 1997. Power Lines, Cancer, and Fear. *New England Journal of Medicine* 337(1): 44–46.

Camus, Albert. 1955. *The Myth of Sisyphus and Other Essays.* Trans. Justin O'Brien. New York: Random House.

C&I. 1999. Monsanto Shifts Focus to Consumer Benefits. Chemistry and Industry 24, 20 December.

Carson, Rachel 1962. *Silent Spring.* Boston: Houghton Mifflin.

Castles, Ian. 1998. The Mismeasure of Nations: A Review Essay on the Human Development Report 1998. *Population and Development Review* 24(4): 831–45.

Catteruccia, F., T. Nolan, T. G. Loukeris, C. Blass, C. Savakis, F. Kafatos, and A. Crisanti. 2000. Stable Germline Transformation of the Malaria Mosquito *Anopheles Stephensi. Nature* 405(6789): 959–62.

CDC. 1999a. *Improvements in Workplace Safety—United States, 1900–1999.* Atlanta: National Institute for Occupational Safety and Health, Centers for Disease Control and Prevention, 11 June.

———. 1999b. CDC on Vaccines and Children's Health: United States 1900–98. *Population and Development Review* 25(2): 391.

———. 1999c. CDC on Infectious Diseases in the United States: 1900–99. *Population and Development Review* 25(3): 635.

———. 1999d. CDC on Infant and Maternal Mortality in the United States: 1900–99. *Population and Development Review* 25(4): 821–26.

———. 2000a. Morbidity and Mortality Weekly Report: Achievements in Public Health, 1900—1999: Family Planning. *JAMA (Journal of the American Medical Association)* 283(3).

———. 2000b. *Morbidity and Mortality Weekly Report:* Achievements in Public Health, 1900–1999: Changes in the Public Health System. *JAMA (Journal of the American Medical Association)* 283(6).

———. 2000c. *Morbidity and Mortality Weekly Report:* Achievements in Public Health, 1900–1999: Fluoridation of Drinking Water to Prevent Dental Caries. *JAMA (Journal of the American Medical Association)* 283(10).

———. 2000d. *Morbidity and Mortality Weekly Report:* Progress Toward Global Poliomyelitis Eradication. *JAMA (Journal of the American Medical Association)* 283(20).

CDHS. 2000a. State Health Director Warns Public about Eating Raw Sprouts. Sacramento: California Department of Health Services, News release, 11 February.

———. 2000b. Salmonellosis Outbreak Associated with Raw Mung Bean Sprouts. Sacramento: California Department of Health Services, News release, 19 April.

———. 2000c. Consumers Warned Not to Drink Unpasteurized Juices Produced by California Day-Fresh Foods. Sacramento: California Department of Health Services, News release, 20 April.

Chambers, Judith. 1999. Discussion. In *Biotechnology and Biosafety: A Forum*, edited by Ismail Serageldin and Wanda W. Collins. Washington, D.C.: World Bank.

Chattopadhyaya, Debiprasad. 1986. *History of Science and Technology in Ancient India: The Beginnings*. Calcutta: Firma KLM Pt. Ltd.

Cheney, Peter, and Krista Foss. 2000. Is Bottled Water Really Better? Tests Show Some Brands Contain Bacteria. *Globe and Mail (Toronto)*, 17 June.

Childe, V. Gordon. 1951. *Man Makes Himself*. New York: Mentor Books.

Christensen, Ole. 1999. Personal Email from Dr. Ole W. Christensen, OCP Liaison Office, WHO, Geneva, Switzerland, 16 February.

Chui, Glennda. 1988. Activists Beset UC, Stanford Labs and a Lab in the Line of Fire. *Science* 239(4845): 1229–32.

Clausen, M., R. Krauter, G. Schachermayr, I. Potrykus, and C. Sautter. 2000. Antifungal Activity of a Virally Encoded Gene in Transgenic Wheat. *Nature Biotechnology* 18(4): 446–49.

CNS. 1999a. Two Leading Researchers Take Issue with Three Recent Studies on the Effects of Genetically Engineered Crops, Cornell University. Cornell News Service, 10 September.

———. 1999b. "Bioavailability" Is the Real Test for DDT Hazard. Cornell News Service, 15 November.

Coates, Craig J. 2000. Malaria: A Mosquito Transformed. *Nature* 405(6789): 900–901.

CofE. 1999. The Church of England's View of Genetically Modified Organisms, http: //www.cofe.anglican.org/view/environ.hyml.

Coffman, Bob. 2000. Biotech Scientists Promote Responsibility, *AgWeb.com* online, 29 March.

Cohen, Mark N. 1987. The Significance of Long-term Changes in Human Diet and Food Economy. In *Food and Evolution: Toward a Theory of Human Food Habits,* edited by Marvin Harris and Eric B. Ross. Philadelphia: Temple University Press.

———. 1989. *Health and the Rise of Civilization*. New Haven: Yale University Press.

———. 1992. The Epidemiology of Civilization. In *Human Impact on the Environment: Ancient Roots, Current Challenges*, edited by Judithe Jacobsen and John Firor. Boulder: Westview Press.

ComEurCom. 2000. Communication from the Commission on the Precautionary Principle. Brussels: Commission of the European Communities, 2 February.

Conko, Gregory, and Fred L. Smith, Jr. 1999. Biotechnology and the Value of Ideas in Escaping the Malthusian Trap. *AgBioForum* 2(3, 4): 150–54.

Connole, Patrick. 1998. Treaty Seeks to Limit Trade in Hazardous Chemicals. *Reuters News Service, Nando.net* online, 4 September.

Conway, Gordon. 2000. Genetically Modified Crops: Risks and Promises. *Conservation Ecology* 4(1).

Cook-Mozaffari, P., Richard Doll, and Leo J. Kinlen. 1993. Fluoridation of Drinking Water. No Evidence of Increased Risk of Cancer. *BMJ (British Medical Journal)* 386(6900): 386.

Cooter, Roger. 2000. Review of Robert N. Proctor, *The Nazi War on Cancer. BMJ (British Medical Journal)* 320(7236): 721.

Copp, Newton, and Andrew Zanella. 1993. *Discovery, Innovation, and Risk: Case Studies in Science and Technology*. Cambridge: MIT Press.

Coulston, F. 1985. Reconsideration of the Dilemma of DDT for the Establishment of an Acceptable Daily Risk Intake. *Regulatory Toxicology and Pharmacology* 5: 332–82.

Cowan, Ruth Schwartz. 1997. *A Social History of American Technology*. New York: Oxford University Press.

CR. 1989. Alar: Not Gone, But Forgotten, and Alar in Apples: Facts and Fantasies. *Consumer Reports* 54(5): 288–91.

———. 1990. Hold the Mold: Aflatoxin in Peanut Butter. *Consumer Reports* 55(9): 591.

Crenson, Matt. 1997. Thousands of Children Jeopardized by Pesticide Use. Associated Press, Nando.net online, 18 December.

Crimins, Eileen M., Yasuhiko Saito, and Dominique Ingegneri. 1997. Trends in Disability-Free Life Expectancy in the U.S. *Population and Development* 23(3): 555–72.

Crissman, J. W. , P. S. Cooke, R. A. Hess, M. S. Marty, and A. B. Liberacki. 2000. Postulated Human Sperm Count Decline May Involve Historic Elimination of Juvenile Iodine Deficiency: A New Hypothesis with Experimental Evidence in the Rat. *Toxicological Sciences* 53(2): 400–410.

Croft, Ashley. 2000. Malaria: Prevention in Travellers. *BMJ (British Medical Journal)* 321(7254): 154–60.

Crosby, Alfred W. 1993. Smallpox. In *The Cambridge World History of Human Disease*, edited by Kenneth F. Kiple. Cambridge: Cambridge University Press.

Crystal, Ben. 2000. Light Fantastic. *New Scientist* 165(2220): 14–17.

Crystal, David. 1987. *The Cambridge Encyclopedia of Language*. Cambridge: Cambridge University Press.

CU. 2000. Letter. Consumer's Union's Reply to Author's Inquiry Concerning Organic Food Testing, 18 April.

Cushing, Frank Hamilton. 1892. Manual Concepts: A Study of the Influence of Hand-usage on Culture-growth. *American Anthropologist* 5(4): 289–317.

David, Christina C., and Keijiro Otsuka. 1994. Modern Rice Technology: Emerging Views and Policy Implications. In *Modern Rice Technology and Income Distribution in Asia*, edited by Christina C. David and Keijiro Otsuka. Boulder: Lynne Reinner Publishers.

Davies, John B. 1994. Sixty Years of Onchocerciasis Vector Control: A Chronological Summary with Comments on Eradication, Reinvasion, and Insecticide Resistance. *Annual Review of Entomology* 39: 23–46.

Davies, Robertson. 1985. *What's Bred in the Bone*. New York: Viking.

Dawkins, Richard. 1985. Sociobiology: The Debate Continues. A Review of *Not in Our Genes: Biology, Ideology and Human Nature* by Steven Rose, Leon J. Kamin, and R. C. Lewontin. New York: Pantheon Books. *New Scientist*, 24 January.

Deacon, Terrence William. 1997. *The Symbolic Species: The Co-evolution of Language and the Brain*. New York: W. W. Norton.

Dean, Malcolm. 2000. At Last, the Fight against Lymphatic Filiariasis Begins. *Lancet* 355(9201).

Decker, Hannah. 2000. Review of Wolfgang Weyers, *Death of Medicine in Nazi Germany: Dermatology and Dermatopathology under the Swastika*. *JAMA (Journal of the American Medical Association)* 283(11).

DeGregori, Thomas R. 1977. Technology and Ceremonial Behavior: Some Aspects of Institutionalism. *Journal of Economic Issues* 11(4): 861–70.

———. 1985. *A Theory of Technology: Continuity and Change in Human Development*. Ames: Iowa State University Press.

———. 1987a. Resources Are Not; They Become. *Journal of Economic Issues* 20(2): 463–70.

———. 1987b. Technology in Our Changing World. Symposium on Science and Technology for Development: Prospects Entering the 21st Century. Washington, D.C.: National Academy of Sciences and U.S. AID.

———. 1988. Population Growth, Resources and Pollution. In *Probleme und Chancen Demographischer Entwicklung in der Dritten Welt,* edited by Gunter Steinmann. Berlin: Springer-Verlag.

———. 1996. Technology Transfer, Economic Development, and the Perpetuation of Poverty: Resource Creation Versus Frugality. In *The Institutional Economics of the International Economy*, edited by John Adams and Anthony Scaperlanda. Boston: Kluwer Academic Publishers.

———. 1998. Technological Progressivism: Guilty as Charged. *Journal of Economic Issues* 32(3): 848–56.

———. 1999. It Has Been a Very Good Century, But. *Priorities for Long Life and Good Health* 11(3): 7–11, 40.

———. 2000a. The Sky Is Not Falling: The Culture of Baseless Fears. *Priorities for Long Life and Good Health* 12(1).

———. 2000b. Let Us Spray: Malaria and DDT in Mozambique. *drkoop.com* online and *ACSH.com* online.

———. 2000c. Sassower's Technoscientific Angst: Ethics+Responsibility: A Review Essay. In *Research in the History of Economic Thought and Methodology*, edited by Warren J. Samuels and Jeff Biddle. Greenwich: JAI Press.

DeGregori, Thomas R., and Randal J. Thompson. 1991. Technology as Domination. *Journal of Scientific and Industrial Research* 50(10): 718–26.

Deichmann, Ute. 2000. An Unholy Alliance: The Nazis Showed That "Politically Responsible" Science Risks Losing Its Soul. *Nature* 405(6788): 739.

della-Cioppa, Guy, and Mary Callan. 2000. Sex, Lies and Herbicides. *Nature Biotechnology* 18(3): 241.

DellaPenna, Dean. 1999. Nutritional Genomics: Manipulating Plant Micronutrients to Improve Human Health. *Science* 285(5426): 375–79.

DeLong, J. Bradford. 1991–2000. The Economic History of the Twentieth Century: Slouching Towards Utopia. online, http://www.j-bradford-delong.net/.

de Macedo, Carlyle Guerra. 1993. Balancing Microbial and Chemical Risks in Disinfection of Drinking Water: The Pan American Perspective. *Bulletin of the Pan American Health Organization* 27(2): 197–200.

Demos, John. 1970. *A Little Commonwealth: Family Life in Plymouth Colony.* New York: Oxford University Press.

Denby, David. 1996. My Problem with Perfection. *New Yorker* 72 (24&25): 64–83, Special Music Issue, 26 August and 2 September.

Desowitz, Robert S. 1991. *The Malaria Capers: More Tales of Parasites and People, Research and Reality.* New York: W.W. Norton.

———. 1993. *The Malaria Capers: More Tales of Parasites and People, Research and Reality.* New York: W.W. Norton.

———. 1997. *Tropical Diseases: From 50,000 BC to 2500 AD.* London: HarperCollins.

Devraj, Ranjit. 2000. Cyclone Victims Are Guinea Pigs for Mutant Food. *Inter Press Service atimes.com* online, 13 June.

Dewey, John. 1929. *The Quest for Certainty: A Study of the Relation of Knowledge and Action.* 1980 reprint. New York: Capricorn Books, G. P. Putnam & Sons.

———. 1934. *Art as Experience.* New York: Minton, Balch & Co.

———. 1938. *Logic: The Theory of Inquiry.* New York: Henry Holt.

———. 1939. *Theory of Valuation. International Encyclopedia of Unified Science.* Foundations of the Unity of Science, 2 (4). Chicago: University of Chicago Press.

———. 1958. *Experience and Nature.* New York: Dover Publications.

DeWitt, Natalie. 2000a. Excising Selectable Markers from Transgenic Plants. *Nature Biotechnology* 18(4): 370.

———. 2000b. Boosting Plant Phosphate Uptake. *Nature Biotechnology* 18(4): 370.

Dickinson, Heather, and Louise Parker. 1999. Quantifying the Effect of Population Mixing on Childhood Leukemia Risk: The Seascale Cluster. *British Journal of Cancer* 81(1): 144–51.

Dilworth, Leah. 1996. *Imagining Indians in the Southwest: Persistent Visions of a Primitive Past*. Washington, D.C.: Smithsonian Institution Press.

DiMatteo, Katherine T. 1997. Does Organic Gardening Foster Foodborne Pathogens. *JAMA (Journal of the American Medical Association)* 277(21).

Dixon, Bernard. 1999. The Paradoxes of Genetically Modified Foods: A Climate of Mistrust Is Obscuring the Many Different Facets of Genetic Modification. *BMJ (British Medical Journal)* 318(7183): 547–88.

Dobelle, William H. 2000. Artificial Vision for the Blind by Connecting a Television Camera to the Visual Cortex. *ASAIO Journal (Journal of the American Society of Artificial Internal Organs)* 46(1).

Doll, Richard. 1999. The Seascale Cluster: A Probable Explanation. *British Journal of Cancer* 81(1).

Doll, Richard, and Richard Peto. 1981. The Causes of Cancer: Quantitative Estimates of Avoidable Risks of Cancer in the United States Today. *Journal of the National Cancer Institute* 66(6): 1191–1308.

Dominick, Raymond H., III. 1992. *The Environmental Movement in Germany: Prophets and Pioneers, 1871–1971*. Bloomington: Indiana University Press.

Dowell, Scott F., Benjamin A. Kupronis, Elizabeth R. Zell, and David K. Shay. 2000. Mortality from Pneumonia in Children in the United States, 1939 through 1996. *New England Journal of Medicine* 342(19): 1399–1407.

Drew, Christopher, and Pam Belluck. 1998. Deadly Bacteria a New Threat to Fruit and Produce in U.S. *New York Times,* 4 January.

Drotman, D. Peter. 2000. Review of Robert W. Proctor, *The Nazi War on Cancer. JAMA (Journal of the American Medical Association)* 283(11).

Drury, William H. 1998. *Chance and Change: Ecology for Conservationists*. Berkeley: University of California Press.

Dyson, Freeman J. 1997. *Imagined Worlds*. Cambridge: Harvard University Press.

———. 1999. Miracles of Rare Device: New Tools, Not New Ideas, Will Usher in Tomorrow's Scientific Marvels. *Sciences* 39(2).

Easley, Michael W. 1997. Please Do Drink the Water. *ACSH Media Update*, 9, August.

Easterbrook, Gregg. 1999a. America the O.K.: Why Life in the U.S. Has Never Been Better. *New Republic* 220(1, 2): 19–25.

———. 1999b. Science Fiction. *New Republic* online, 220(8), 30 August.

Easterlin, Richard A. 2000. The Worldwide Standard of Living Since 1800. *Journal of Economic Perspectives.* 14(1): 7–26.

Economist. 1994. Power to the People: A Survey of Energy. *Economist* 331(7868): 1–18.

EID. 1997. *Emerging Infectious Diseases Journal* 3(4), October–December. Special Issue—The National Conference on Emerging Foodborne

Pathogens: Implications and Control, March 24–26, 1997, Alexandria, Va.

Eisenberg, David M. , Roger B. Davis, Susan L. Ettner, Scott Appel, Sonja Wilkey, Maria Van Rompay, and Ronald C. Kessler. 1998. Abstracts: Trends in Alternative Medicine Use in the United States, 1990–1997: Results of a Follow-up National Survey. *JAMA (Journal of the American Medical Association)* 280(18): 1569.

Eisenstein, Elizabeth L. 1979. *The Printing Press as an Agent of Change.* Cambridge: Cambridge University Press.

———. 1983. *The Printing Revolution in Early Modern Europe.* Cambridge: Cambridge University Press.

Eizenstat, Stuart E. 1999. Personal View: Why We Should Welcome Biotechnology. *Financial Times (London)*, 16 April.

Ekperigin, H. E. 2000. Use of Poultry Litter or Manure as a Fertilizer for Croplands: Impact on Animal and Public Health. *FDA Veterinarian* 15(3).

Elder, J., and L. Cooley. 1995. *Sustainable Settlement and Development of the Onchocerciasis Control Programme Area: Proceedings of a Ministerial Meeting.* Technical Paper, 310. Washington, D.C.: World Bank.

Eliot, T. S. 1952. Little Gidding, Last of the Four Quartets. *The Complete Poems and Plays, 1909–1950.* New York: Harcourt, Brace and World, Inc.

Emsley, John. 1994. *The Consumer's Good Chemical Guide: A Jargon-Free Guide to the Chemicals of Everyday Life.* Oxford: W. H. Freeman and Co.

Enserink, Martin. 1999a. Bioengineering: Preliminary Data Touch Off Genetic Food Fight. *Science* 283(5405): 1094–95.

———. 1999b. Transgenic Food Debate: *The Lancet* Scolded over Pusztai Paper. *Science* 286(5440): 656.

———. 1999c. Industry Response: Ag Biotech Moves to Mollify Its Critics. *Science* 286(5445).

Ernest, E. 1999. Second Thoughts about Safety of St John's Wort. *Lancet* 354(9195): 2014–16.

Ernst, Pierre, and Yvon Cormier. 2000. Relative Scarcity of Asthma and Atopy among Rural Adolescents Raised on a Farm. *American Journal of Respiratory and Critical Care Medicine* 161(5): 1563–66.

ERS. 2000. Does Genetic Engineering Reduce Crop Pesticide Use? Washington, D.C.: Economic Research Service, ERS-AO-273, U. S. Department of Agriculture, August.

Etheridge, Elizabeth W. 1993. Pellagra. In *The Cambridge World History of Human Disease,* edited by Kenneth F. Kiple. Cambridge: Cambridge University Press.

EW (Environment Writer). 1997. DDT ($C_{14}H_9C_{l5}$)/DDE ($C_{14}H_8C_{l4}$)/DDD ($C_{13}H_{10}C_{l4}$): Chemical Backgrounder. Washington, D.C.: Environmental Health Center, Division of the National Safety Council online.

Ewald, Paul W. 1994. *Evolution of Infectious Disease*. New York: Oxford University Press.

Ewen, Stanley W. B., and Arpad Pusztai. 1999a. Health Risks of Genetically Modified Foods. *Lancet* 354(9179).

———. 1999b. Effect of Diets Containing Genetically Modified Potatoes Expressing Galanthus Nivalis Lectin on Rat Small Intestine. *Lancet* 354(9187): 1353.

Fagin, Dan. 2000. Not a Cancer Cause? Breast Cancer Study: No Link to DDT, 3 Other Chemicals. *Newsday*, 14 November.

Falck-Zepeda, J. B. , G. Traxler, and R. G. Nelson. 1999. *Rent Creation and Distribution from the First Three Years of Planting Bt Cotton*. ISAAA (International Service for the Acquisition of Agri-Biotech Applications) Briefs No. 14-1999.

———. 2000a. Surplus Distribution from the Introduction of a Biotechnology Innovation. *American Journal of Agricultural Economics* 82(2): 360–69.

———. 2000b. Rent Creation and Distribution from Biotechnology Innovations: The Case of Bt Cotton and Herbicide-Tolerant Soybeans in 1997. *Agribusiness* 16(1): 21–32.

FAO. 1989. *Report of the Commission on Plant Genetic Resources*. Rome: Food and Agriculture Organization of the United Nations.

———. 1990. *International Code of Conduct on the Distribution and Use of Pesticides (Amended to Include Prior Informed Consent in Article 9 as Adopted by the 25th Session of the FAO Conference in November 1989)*. Rome: Food and Agriculture Organization of the United Nations.

———. 1996. Biotechnology and Food Safety: Report of a Joint FAO/WHO Consultation, Rome, Italy, 30 September–4 October 1996. FAO Food and Nutrition Paper, No. 61 0254-4725. Rome: Food and Agriculture Organization of the United Nations.

———. 1999. *The State of Food Insecurity in the World, 1999*. Rome: Food and Agriculture Organization of the United Nations.

Farah, Jumanah. 1994. Pesticide Policies in Developing Countries: Do They Encourage Excessive Use? World Bank Discussion Paper No. 238. Washington, D.C.: World Bank.

Farland, William H., and Herman J. Gibb. 1993. U.S. Perspective on Balancing Chemical and Microbial Risks of Disinfection. In *Safety of Water Disinfection: Balancing Chemical and Microbial Risks,* edited by Gunther F. Cruan. Washington, D.C.: ILSI (International Life Sciences Institute) Press.

Farrington, Benjamin. 1944. *Greek Science*. London: Penguin Books.

———. 1947. *Head and Hand in Ancient Greece*. London: Studies in the Social Relations of Thought.

Faulkner, William. 1956. *Faulkner at Nagano*. Edited by Robert A. Jellife. Tokyo: Kenkyusha.

Feldbaum, Carl. 1999. One UK Biotech Company's Demise Shows Us All What We Stand to Lose. *Financial Times (London),* 16 September.

Felsot, Allan S. 2000a. Insecticidal Genes: Part I: From Tight Fit to Uptight. *Agrichemical and Environmental News* 167 (March).

———. 2000b. Insecticidal Genes: Part 2: Human Health Hoopla. *Agrichemical and Environmental News* 168 (April).

———. 2000c. Insecticidal Genes: Part 3: Long Live the Monarch. *Agrichemical and Environmental News* 169 (May).

Fenner, Frank. 1996. History of Smallpox. In *Microbe Hunters—Then and Now,* edited by H. Koprowski and Michael B. A. Oldstone. Bloomington: Med-Ed Press.

Fenner, Frank, D. A. Henderson, I. Arita, Z. Jezek, and I. D. Ladnyi. 1988. *Smallpox and Its Eradication.* Geneva: World Health Organization.

Ferguson, Eugene S. 1977. The Mind's Eye: Nonverbal Thought in Technology. *Science* 197(4306): 827–36.

———. 1978. Elegant Inventions: The Artistic Component of Technology. *Technology and Culture* 19(3): 450–60.

———. 1979. The American-ness of American Technology. *Technology and Culture* 20(1): 3–24

———. 1992. *Engineering and the Mind's Eye.* Cambridge: MIT Press.

Ferguson, Kirsty. 1997. Steiner's Philosophy on Compost: The Plot Thickens. *The Independent (London),* 1 November.

Fett, William F. 2000. Naturally Occurring Biofilms on Alfalfa and Other Types of Sprouts. *Journal of Food Protection* 63(5): 625–32.

Fischer, P., R. Garms, D. W. Buttner, W. Kipp, J. Bamuhiiga, and J. Yocha. 1997. Reduced Prevalence of Onchocerciasis in Uganda Following Either Deforestation or Vector Control with DDT. *The East African Medical Journal: The Organ of the Medical Association of East Africa (also called Kenya and East African Medical Journal)* 74(5): 321–25.

Flavel, Richard. 1999. Biotechnology and Food and Nutrition Needs. In *Biotechnology for Developing-Country Agriculture: Problems and Opportunities,* edited by Gabrielle J. Persley. Washington, D.C.: International Food Policy Research Institute.

Fleming, D. M., R. Sunderland, K. W. Cross, and A. M. Ross. 2000a. Declining Incidence of Episodes of Asthma: A Study of Trends in New Episodes Presenting to General Practitioners in the Period 1989–98. *Thorax* 55(8): 657–61.

———. 2000b. Comparison of the Seasonal Patterns of Asthma Identified in General Practitioner Episodes, Hospital Admissions, and Deaths. *Thorax* 55(8): 662–65.

Florig, H. Keith. 1992. Containing the Cost of the EMF Problem. *Science* 257(5069): 468–69, 490, 492.

Floud, Roderick. 1990. *Height, Health, and History: Nutritional Status in the United Kingdom, 1750–1980*. Cambridge: Cambridge University Press.

Fodor, J. A. 1983. *Modularity of Mind: An Essay on Faculty Psychology*. Cambridge, Mass.: MIT Press.

Fogel, Robert W., and Dora L. Costa. 1997. A Theory of Technophysio Evolution, with Some Implications for Forecasting Population, Health Care Costs, and Pension Costs. *Demography* 34(1): 49–66.

Foner, Henry. 1993. Wagner, the Nazis, and Animal Rights. *CFAAR Newsletter* 5(1): 10–11.

Foster, Catherine. 1988. Can-do Technology: Breaking Barriers. *Christian Science Monitor,* 26 July.

Foster, Kenneth R., David E. Bernstein, and Peter W. Huber, eds. 1993. *Phantom Risk: Scientific Inference and the Law*. Cambridge: MIT Press.

Foster, Kenneth R., Paolo Vecchia, and Michael H. Repacholi. 2000. Risk management: Science and the Precautionary Principle. *Science* 288(5468): 979–81.

Foster, Max. 1998. Paris Mayor Warns Eiffel Tower May Go American. BBC World Service online, 8 July.

Fox, Robin 1996. State of the Art/science in Anthropology. In *The Flight from Science and Reason,* edited by Paul R. Gross, Norman Levitt, and Martin W. Lewis. New York: The New York Academy of Sciences.

Frary, Anne, T. Clint Nesbitt, Amy Frary, Silvana Grandillo, Esther van der Knaap, Bin Cong, Jiping Liu, Jaroslaw Meller, Ron Elber, Kevin B. Alpert, and Steven D. Tanksley. 2000. fw2. 2: A Quantitative Trait Locus Key to the Volution of Tomato Fruit Size. *Science* 289(5476): 85–88.

Freedman, D. O., T. R. Unnasch, A. Merriweather, and K. Awadzi. 1994. Truly Infection-Free Persons Are Rare in Areas Hyperendemic for African Onchocerciasis. *The Journal of Infectious Diseases: Official publication of the Infectious Diseases Society of America* 170(4): 1054.

French, J. R. 1990. Stalking Celery. *Priorities: For Long Life and Good Health,* Summer, 15–16.

Freyer, Felice J. 1997. Alternative Medicine Has Probably Never Been More Popular. *Providence Journal-Bulletin,* 3 November.

Friedrich, M. J. 1999. Genetically Enhanced Rice to Help Fight Malnutrition. *JAMA (Journal of the American Medical Association)* 282(16).

Frisvold, George, John Sullivan, and Anton Raneses. 1999. Who Gains from Genetic Improvements in U.S. Crops. *AgBioForum* 2(3&4): 226–35.

Frommer, Wolf B., Uwe Ludewig, and Doris Rentsch. 1999. Enhanced: Taking Transgenic Plants with a Pinch of Salt. *Science* 285(5431): 1222–23.

Fugh-Berman, Adriane. 2000. Herb-drug Interactions. *Lancet* 355(9198).

Furedi, Frank. 1997. *Culture of Fear: Risk-taking and the Morality of Low Expectation.* London: Cassell.

Galison, Peter Louis. 1997. *Image and Logic: A Material Culture of Microphysics*. Chicago: University of Chicago Press.

GAO Report. 2000a. *Pesticides: Improvements Needed to Ensure the Safety of Farmworkers and Their Children*. GAO Report, RCED-00-40, March. Washington, D.C.: General Accounting Office.

———. 2000b. *Food Safety: Improvements Needed in Overseeing the Safety of Dietary Supplements and "Functional Foods."* GAO Report, RCED-00-156, 11 July. Washington, D.C.: United States General Accounting Office.

Garate, T., F. J. Conraths, W. Harnett, D. W. Buttner, and R. M. E. Parkhouse. 1996. Identification of *Onchocerca volvulus* as an Antigen Mainly Recognized by Antibodies in Chronic Hyper-reactive Onchodermatitis. *American Journal of Tropical Medicine and Hygiene* 54(5): 490–97.

Garn, Stanley M. 1994. Uses of the Past. *American Journal of Human Biology* 6(1): 89–96.

Gaskell, George, Martin W. Bauer, John Durant, and Nicholas C. Allum. 1999. Worlds Apart? The Reception of Genetically Modified Foods in Europe and the U.S. *Science* 285(5426).

Gasson, M. J. 1999. Genetically Modified Foods Face Rigorous Safety Evaluation. *Nature* 402(6759): 229.

Gawande, Atul. 1999. The Cancer-Cluster Myth: When a Dozen People in a Neighborhood Develop Tumors, It Can't Be a Coincidence, or Can It? *New Yorker* 54(45): 34–37.

Gay, Peter. 1968. *Weimar Culture: The Outsider As Insider*. New York: Harper & Row.

Geertz, Clifford. 1962. The Growth of Culture and the Evolution of Mind. In *Theory of Mind*, edited by J. Scherl. Glencoe: The Free Press of Glencoe; reprinted in Clifford Geertz, (1973), *The Interpretation of Culture: Selected Essays*. New York: Basic Books.

———. 1966. The Impact of the Concept of Culture on the Concept of Man. In *New Views of the Nature of Man*, edited by J. Platt. Chicago: The University of Chicago Press; reprinted in Clifford Geertz (1973), *The Interpretation of Culture: Selected Essays*. New York: Basic Books.

———. 1973. *The Interpretation of Culture: Selected Essays*. New York: Basic Books.

———. 1977. The Transition to Humanity. In *Horizons of Anthropology*, edited by Sol Tax and Leslie G. Freeman. 2d ed. Chicago: Aldine Publishing Co.

———. 1988. *Works and Lives: The Anthropologist as Author*. Stanford: Stanford University Press.

Gelbard, Alene, Carl Haub, and Mary M. Kent. 1999. World Population Beyond Six Billion. *Population Bulletin* 54(1).

Gern, James E., and Scott T. Weiss. 2000. Protection against Atopic Diseases by Measles: A Rash Conclusion? *JAMA (Journal of the American Medical Association)* 283(3).

Gereda, J. E., D. Y. M. Leung, A. Thatayatikom, J. E. Streib, M. R. Price, M. D. Klinnert, and A. H. Liu. 2000. Relation between House-dust Endotoxin Exposure, Type 1 T-cell Development, and Allergen Sensitisation in Infants at High Risk of Asthma. *Lancet* 355(9216): 1680–83.

Gettens, Rutherford J., and George L. Stout. 1966. *Painting Materials: A Short Encyclopedia.* New York: Dover Publications.

GFHR. 2000. *The 10/90 Gap Report on Health Research 2000.* Geneva: Global Forum for Health Research, World Health Organization.

Ghosh, Amitabha. 1994. History of Technology in India: A Critical Review. *SHOT Newsletter: The Society for the History of Technology* (66): 5–7.

Ghosh, Santikumar. 1987. Economies in Human Fulfillment: A Wider Perspective. *Bulletin of the Ramakrishna Mission Institute of Culture* 38(6): 148–53.

Gianessi, Leonard. 1993. The Quixotic Quest for Chemical-free Farming. *Issues in Science and Technology* 10(1): 29–36.

Gibson, Kathleen R. 1993a. Animal Minds, Human Minds. In *Tools, Language, and Cognition in Human Evolution,* edited by Kathleen R. Gibson and Tim Ingold. New York: Cambridge University Press.

———. 1993b. Tool Use, Language, and Social Behavior in Relationship to Information Processing Capacities. In *Tools, Language, and Cognition in Human Evolution,* edited by Kathleen R. Gibson and Tim Ingold. New York: Cambridge University Press.

Gilhooly, Joanne. 2000. Malaria Could Be Europe-bound: *Anopheles gambie* mosquitoes. BBC World Service online, 12 June.

Gilmore, Richard. 2000. Agbiotech and World Food Security: Threat or Boon? *Nature Biotechnology* 18(4): 361.

Gipe, Paul. 1993. The Wind Industry's Experience with Aesthetic Criticism. *Leonardo* 26(3): 243–48.

Giroud, Emmanuel. 2000. Iodine Campaign Halves Retardation Rates. *Electronic Mail & Guardian*, 15 February.

Gladen, B. C., and W. J. Rogan. 1995. DDE and Shortened Duration of Lactation in a Northern Mexican Town. *American Journal of Public Health* 85(4): 504–8.

Gladwell, Malcolm. 1999. Science and the Perils of a Parable. *New Yorker* 54(41): 23–24.

Glynn, A. A. 1989. The Great Egg Debate. *Nature* 339(6227): 671–72.

Goklany, Indur M. 1999. Meeting Global Food Needs: The Environmental Trade-offs between Increasing Land Conversion and Land Productivity. *Technology: Journal of Science Serving Legislative, Regulatory, and Judicial Systems* 6(2-3): 107–30.

———. 2000. Richer is More Resilient: Dealing with Climate Change and More Urgent Environmental Problems. In *Earth Report 2000: Revisiting the True State of the Planet,* edited by Ronald Bailey. New York: McGraw Hill.

Gordon, Robert B. 1988. Strength and Structure of Wrought Iron. *Archeomaterials* 2(2): 109–37.

Gottlieb, Scott. 2000. Chinese Herb May Cause Cancer. *BMJ (British Medical Journal)* 320(7250): 1623.

Goubert, Jean-Pierre. 1989. *The Conquest of Water: The Advent of Health in the Industrial Age.* Translated by Andrew Wilson. Princeton, N. J.: Princeton University Press.

Gowlett, J. A. J. 1992. Early Human Mental Abilities. In *Cambridge Encyclopedia of Human Evolution*, edited by Steve Jones, Robert D. Martin, and David R. Pilbeam. New York: Cambridge University Press.

Graham, John D., and Jonathan Baert Wiener. 1997. Confronting Risk Tradeoffs. In *Risk vs. Risk: Tradeoffs in Protecting Health and the Environment,* edited by John D. Graham and Jonathan Baert Wiener. Cambridge: Harvard University Press.

Gray, George M., and John D. Graham. 1997. Regulating Pesticides. In *Risk vs. Risk: Tradeoffs in Protecting Health and the Environment,* edited by John D. Graham and Jonathan Baert Wiener. Cambridge: Harvard University Press.

Greenpeace. 2000. Panel Made to Study Biotechnology Issues: Controversy over Genetically Altered Foods Won't Be Resolved, Greenpeace Activist Warns. *Globe and Mail,* 19 February.

Greensfelder, Liese. 2000. Alternative Medicine: Herbal Product Linked to Cancer. *Science* 288(5473): 1946.

Gribble, Gordon W. 1999. Chlorine-element from Hell or Gift of God? The Scientific Side of the Chlorine Story. *Technology: Journal of Science Serving Legislative, Regulatory, and Judicial Systems* 6(2-3): 193–201.

Grodstein, Francine, Richard Mayeux, and Meir J. Stampfer. 2000. Tofu and Cognitive Function: Food for Thought. *Journal of the American College of Nutrition* 19(2): 207–9.

Gross, Paul R., and Norman Levitt. 1994. *Higher Superstition: The Academic Left and Its Quarrels with Science.* Baltimore: Johns Hopkins University Press.

Grzimek, Bernhard, and Michael Grzimek. 1961. *Serengeti Shall Not Die.* New York: Dutton.

Gugliotta, Guy. 2000a. Herbal Products Take a Human Toll: Alternative Medicines Promise Health, But Often Don't Deliver. *Washington Post,* 19 March.

———. 2000b. Diet Supplement Marketers Target Kids. *Washington Post,* 18 June.

Guyer, Bernard, Donna L. Hoyert, Joyce A. Martin, Stephanie J. Ventura, Marian F. MacDorman, and Donna M. Strobino. 1999. Annual Summary of Vital Statistics—1998. *Pediatrics* 104(6): 1229–46.

Gwatkin, Davidson R., Michel Guillot, and Patrick Heuveline. 1999. The Burden of Disease among the Global Poor. *Lancet* 354(9178): 586–89.

Hackett, Allan, Indira Nathan, and Lynne Burgess. 1998. Is a Vegetarian Diet Adequate for Children? *Nutrition and Health* 12(3): 189–95.

Hall, Peter. 1998. *Cities in Civilization*. New York: Pantheon Books.

Halverson, John. 1987. Art for Art's Sake in the Paleolithic. *Current Anthropology* 28(1): 63–71, 82–89.

Hamilton, William J., III. 1987. Omnivorous Primate Diet and Human Overconsumption of Meat. In *Food and Evolution: Toward a Theory of Human Food Habits*, edited by Marvin Harris and Eric B. Ross. Philadelphia: Temple University Press.

Hamlin, Christopher. 1990. *A Science of Impurity: Water Analysis in Nineteenth Century Britain*. Berkeley and Los Angeles: University of California Press.

Hansen, Stein. 1992. Population and the Environment. *African Development Review: Special Issue on Population Growth and Sustainable Development in Africa* 4(2): 118–64.

Hapgood, Fred. 2000. Garage Biotech Is Here or Just around the Corner. *Civilization: The Magazine of the Library of Congress* 7(2): 46–51.

Hart, Arthur A. 1977. An Interpretive Case Study: Custer, Idaho. In *Ghost Towns and Mining Camps: Selected Papers 25–28*. Washington, D.C.: Preservation Press for the National Trust for Historic Preservation.

Harris, W. V. 1989. *Ancient Literacy*. Cambridge, Mass.: Harvard University Press.

Harten, A. M. van. 1998. *Mutation Breeding: Theory and Practical Applications*. New York: Cambridge University Press.

Havelock, Eric A. 1982. *The Literate Revolution in Greece and Its Cultural Consequences*. Princeton: Princeton University Press.

Hawking, Stephen W. 1988. *A Brief History of Time: From the Big Bang to Black Holes*. New York: Bantam Books.

Hayden, Brian. 1981. Research and Development in the Stone Age: Technological Transitions Among Hunter-gatherers. *Current Anthropology* 22(5): 519–48.

Heath, Clark W., Jr. 1997. Pesticides and Cancer Risk. *Cancer (American Cancer Society)* 80(10): 1887–88.

Helmuth, Laura. 2000. Both Sides Claim Victory in Trade Pact. *Science* 287(5454): 782–83.

Helzlsouer, K. J., J. A. Alberg, H. Y. Huang, S. C. Hoffman, P. T. Strickland, J. W. Brock, V. W. Burse, L. L. Needham, D. A. Bell, J. A. Lavigne, J. D. Yager, and G. W. Comstock. 1999. Serum Concentrations of Organochlorine Compounds and the Subsequent Development of Breast Cancer. *Cancer Epidemiology Biomarkers & Prevention* 8: 525–32.

Henderson, Donald A. 1994. Health Issues. In *Poverty Policy in Developing Countries,* edited by Thomas R. DeGregori and Harrell R. Rodgers, Jr. Greenwich: JAI Press.

———. 1996. Smallpox Eradication. In *Microbe Hunters-Then and Now*, edited by H. Koprowski and Michael B. A. Oldstone. Bloomington: Med-Ed Press.

Henry, Sara Hale, F. Xavier Bosch, Terry C. Troxell, and P. Michael Bolger. 1999. Reducing Liver Cancer: Global Control of Aflatoxin. *Science* 286(5449): 2453–54.

Herbert, Victor. 1992. L-tryptophan: A Medicolegal Case against Over-the-counter Marketing of Supplements of Amino Acids. *Nutrition Today* 27(2): 27–30.

Herman, Robin. 1998. Experts Pondering Implications of Rising Life Expectancy. *Houston Chronicle Washington Post News*, 27 May.

Heuveline, Patrick. 1999. The Global and Regional Impact of Mortality and Fertility Transitions, 1950–2000. *Population and Development Review* 25(4): 681–702.

Hill, Matthew. 1999. *Vegetarian Diet Linked to Genital Defects: A Vegetarian Diet May Put Babies at Risk*. BBC World Service online, 25 February.

Hillman, J. R. 2000. *Report of the Director*. SCRI (Scottish Crop Research Institute) 1998/99 Annual Report 11-44 Invergowie, Dundee, Scotland, January.

Hindle, Brooke. 1983. *Emulation and Invention*. New York: W.W. Norton.

Hiscoe, Helen B. 1983. Does Being Natural Make It Good? *New England Journal of Medicine* 308(24): 1474.

Hoban, Thomas J. 1999. International Acceptance of Agricultural Biotechnology. In *Proceedings of the Annual Meeting of the National Agricultural Biotechnology Council*. Ithaca: National Agricultural Biotechnology Council.

———. 2000. Social Controversy and Consumer Acceptance of Agricultural Biotechnology. *Journal of BioLaw and Business* 3(3).

Hodgson, John. 2000a. GM Health Food. *Nature Biotechnology* 18(3): 247.

———. 2000b. GMO Roundup. *Nature Biotechnology* 18(4).

Hoerauf, Achim, Lars Volkmann, Christoph Hamelmann, Ohene Adjei, Ingo B. Autenrieth, Bernhard Fleischer, and Dietrich W. Buttner. 2000. Endosymbiotic Bacteria in Worms as Targets for a Novel Chemotherapy in Filariasis. *Lancet* 355(9211): 1242–43.

HOL. 1998. *EC Regulations of Genetic Modification in Agriculture*. House of Lords, Select Committee on the European Communities, Session 1998–1999, 2d report.

Holland, Bernard. 1999. Old Records in High Tech and Low. *New York Times*, 10 April.

Horiuchi, Shiro. 2000. Demography: Greater Lifetime Expectations. *Nature* 405(6788): 744–45.

Horton, Richard. 1999. Genetically Modified Foods: "Absurd" Concern or Welcome Dialogue? *Lancet* 354(9187): 1312.

Hougard, J.-M., H. Agoua, L. Yameogo, K. L. B. Akpoboua, A. Seketeli, and K. Y. Dadzie. 1998. Blackfly Control: What Choices after Onchocerciasis? *World Health Forum* 19(3): 281–84.

Hougard, J.-M., L. Yameogo, A. Seketeli, B. Boatin, and K. Y. Dadzie. 1997. Twenty-two Years of Blackfly Control in the Onchocerciasis Control Programme in West Africa. *Parasitology Today* 13(11): 425–31.

Hougard, J.-M., P. Poudiougo, P. Guillet, C. Back, L. K. Akpoboua, and D. Quillevere. 1993. Criteria for the Selection of Larvicides by the Onchocerciasis Control Programme in West Africa. *Annals of Tropical Medicine and Parasitology* 87(5): 435–42.

Howard, Seymour. 1985. The Steel Pen and the Modern Line of Beauty. *Technology and Culture* 26(4): 785–98.

HRW. 2000. *Fingers to the Bone: United States Failure to Protect Child Farmworkers*. New York: Human Rights Watch.

Huckel, John F. 1913 and 1934. *First Families of the Southwest*. Kansas City: Fred Harvey.

Hulka, Barbara S., Nancy L. Kerkvliet, and Peter Tugwell. 2000. Experience of a Scientific Panel Formed to Advise the Federal Judiciary on Silicone Breast Implants. *New England Journal of Medicine* 342(11).

Hunter, David, Susan E. Hankinson, Francine Laden, Graham A. Colditz, JoAnn E. Manson, Walter C. Willett, Frank E. Speizer, and Mary S. Wolff. 1997. Plasma Organochlorine Levels and the Risk of Breast Cancer. *New England Journal of Medicine* 337(18): 1253–58.

Hu Shih. 1928. The Civilizations of East and West. In *Whither Mankind: A Panorama of Modern Civilization,* edited by Charles Beard. London: Longmans, Green, and Co.

IAEA. 2000. Radiological Consequences of Chernobyl Accident. UN Scientific Committee on Effects of Atomic Radiation Confirms Earlier IAEA Assessments, International Atomic Energy Agency Press release, 13 June.

IARC. 1987. An Updating of IARC Monographs volumes 1 to 42., Supplement 7 IARC Monographs on the Evaluation of Carcinogenic Risks to Humans: Overall Evaluations of Carcinogenicity. Geneva: World Health Organization, International Agency for Research on Cancer, 186–89, 322–26.

———. 1991. Occupational Exposures in Insecticide Application, and Some Pesticides. IARC Monographs on the Evaluation of Carcinogenic Risks to Humans. Geneva: World Health Organization, International Agency for Research on Cancer.

IFPRI. 1999. Pushing Back Poverty in India. *News & Views: A 2020 Vision for Food, Agriculture, and the Environment*. International Food Policy Research Institute, September.

IFRCRCS. 2000a. Diseases Are Turning into Silent Disasters Warns Red Cross/Red Crescent Disaster Report. Geneva: International Federation of Red Cross and Red Crescent Societies, Press release, 28 June.

————. 2000b. *World Disasters Report 2000: Focus on Public Health.* Geneva: International Federation of Red Cross and Red Crescent Societies, June

Inglehart, Ronald. 1997. *Modernization and Postmodernization: Cultural, Economic, and Political Change in 43 Societies.* Princeton: Princeton University Press.

Ingold, Tim. 1993. Tool Use, Sociality, and Intelligence. In *Tools, Language, and Cognition in Human Evolution*, edited by Kathleen R. Gibson and Tim Ingold. New York: Cambridge University Press.

IOM. 2000a. *Clearing the Air: Asthma and Indoor Air Exposures.* Washington, D.C.: National Academy of Sciences, Institute of Medicine, Institute of Medicine, Committee on the Assessment of Asthma and Indoor Air.

————. 2000b. Dietary Reference Intakes for Vitamin C, Vitamin E, Selenium, and Carotenoids. Washington, D.C.: National Academy of Sciences, Subcommittees on Upper Reference Levels of Nutrients and Interpretation and Uses of Dietary Reference Intakes, Standing Committee on the Scientific Evaluation of Dietary Reference Intakes, Food and Nutrition Board.

————. 2000c. *Toward Environmental Justice: Research, Education, and Health Policy Needs.* Washington, D.C.: National Academy of Sciences, Institute of Medicine, National Academy Press.

IPMnet. 2000. Synopsis of Selected Pest Management Research. *IPMnet News* (Consortium for International Crop Protection) (77), May.

Irwin, David. 1972. Industry Plunders Antiquity. *Apollo* 96(128): 288–97 (new series).

————. 1976. The Industrial Revolution and the Dissemination of Neo-classical Taste. *Studies on Voltaire and the Eighteenth Century* 153: 1087–98.

Jackson, John Brinckerhoff. 1984. *Discovering the Vernacular Landscapes.* New Haven: Yale University Press.

Jacobs, Jennifer, Edward H. Chapman, and Dean Crothers. 1998. Patient Characteristics and Practice Patterns of Physicians Using Homeopathy. *Archives of Family Medicine* 7(7): 537–40.

Jaeger, Rudolph, and Arlene L. Weiss. 1993. Trichloroethylene: Toxicology and Epidemiology: A Critical Review of the Literature. In *Phantom Risk: Scientific Inference and the Law*, edited by Kenneth R. Foster, David E. Bernstein, and Peter W. Huber. Cambridge: MIT Press.

Jaffa, Kent C., Han Kim, and Tim E. Aldrich. 2000. Relative Merits of Contemporary Measurements and Historical Calculated Fields in the Swedish Childhood Cancer Study. *Epidemiology* 11(3).

JAMA. 1997. Medical News and Perspectives. *JAMA (Journal of the American Medical Association)* 277(2): 97–98.

———. 1998. Alternative Medicine Theme Issue. *JAMA (Journal of the American Medical Association)* 280(18).

———. 1999a. Alfalfa Sprout Seeds Seen as Source of Salmonella Outbreaks: Current Sprouting Methods Inadequate to Prevent Such Events. *JAMA (Journal of the American Medical Association)* 281(2).

———. 1999b. Outbreak of Salmonella Serotype Muenchen Infections Associated with Unpasteurized Orange Juice: United States and Canada, June 1999. *JAMA (Journal of the American Medical Association)* 282(8): 726.

James, Clive, and Anatole Krattiger. 1999. The Role of the Private Sector. In *Biotechnology for Developing-Country Agriculture: Problems and Opportunities,* edited by Gabrielle J. Persley. Washington, D.C.: International Food Policy Research Institute.

Janowsky, Esther C., Lawrence L. Kupper, and Barbara S. Hulka. 2000. Meta-Analyses of the Relation between Silicone Breast Implants and the Risk of Connective-Tissue Diseases. *New England Journal of Medicine* 342(11): 781–90.

Jayarsman, K. S. 2000a. Funding Crisis for Indian Biotech Center. *Nature* 403(6671): 694.

———. 2000b. . . . as India Pushes Ahead with Plant Database. *Nature* 405(6784): 267.

———. 2000c. GM Food "Dumped on India as Food Aid." *Nature* 405(6789): 875.

Jeffers, Robinson. 1941. The Answer. In *The Selected Poetry of Robinson Jeffers*. New York: Random House.

Jenkins, Holman W., Jr. 1999. Fun Facts to Know and Tell about Biotechnology. *Wall Street Journal,* 17 November.

JIC. 1999. "Green Revolution" Gene Isolated by John Innes Scientists. Norwich: John Innes Centre Press release, 12 July.

Jobst, K. A., M. McIntyre, D. St. George, M. Whitelegg, P. A. G. M. DeSmet, D. J. Touw, and D. Wheatley. 2000. Safety of St. John's Wort (Hypericum perforatum): Three Letters. *Lancet* 355(9203): 575.

Johnson, D. Gale. 2000. Population Food and Knowledge. *American Economic Review* 90(1): 1–14.

Johnson, Pamela, and Ann Van Dusen. 1994. Poverty Policies: The Role of Research: Lessons from the Health Sector. In *Poverty Policy in Developing Countries,* edited by Thomas R. DeGregori and Harrell R. Rodgers. Greenwich: JAI Press.

Johnston, Julianne. 2000. Group of 1,200 Scientists Endorse Biotech Crops at Annual Meeting. The Society for In Vitro Biology *Agweb.com* online, 15 June.

Jones, Timothy F., Allen S. Craig, Debbie Hoy, Elaine W. Gunter, David L. Ashley, Dana B. Barr, John W. Brock, and William Schaffner. 2000. Mass Psychogenic Illness Attributed to Toxic Exposure at a High School. *New England Journal of Medicine* 342(2).

Jukes, Thomas H. 1974. Insecticides in Health, Agriculture, and the Environment. *Naturwissenschaften* 61: 6–16.

Juma, Calestous. 1989. *The Gene Hunters: Biotechnology and the Scramble for Seeds*. Princeton: Princeton University Press.

———. 2000. Science, Technology, and Economic Growth: Africa's Biopolicy Agenda in the 21st Century. United Nations, Institute for Natural Resources in Africa, UNU/INRA Annual Lectures, 1999.

Kamat, S. R. 1998. Air Pollution and Respiratory Problems. In *Respiratory Medicine in the Tropics,* edited by J. N. Pande. New Delhi: Oxford University Press.

Kaufman, Ron. 1993. Unprecedented Justice Department Report Documents Extent of Animal Activists Attacks on Research Labs. *Scientist: The Newspaper for the Science Professional* 7(20): 1, 6.

Keane, F. M., S. E. Munn, A. W. P. du Vivier, N. F. Taylor, and E. M. Higgins. 1999. Analysis of Chinese Herbal Creams Prescribed for Dermatological Conditions. *BMJ (British Medical Journal)* 318(7183): 563–64.

Kedzierski, Marie. 1992. Vaccines and Immunization. Inside Science No. 53, Special Section. *New Scientist* 133(1807).

Kermode, Frank. 1985. Apocalypse and the Modern. In *Vision of Apocalypse: End or Rebirth?* edited by Saul Friedlander, Gerald Holton, Leo Marx,, and Eugene Skolnikoff. New York and London: Holmes and Meier.

Kim, A., and B. Benton. 1995. Cost-Benefit Analysis of the Onchocerciasis Control Program (OCP). Technical Paper No. 282. Washington, D.C.: World Bank.

Kinlen, Leo J. 1988a. Evidence for an Infective Cause of Childhood Leukaemia: Comparison of a Scottish New Town with Nuclear Reprocessing Sites in Britain. *Lancet* 2(8624).

———. 1988b. The Longitudinal Study and the Social Distribution of Cancer. *BMJ (British Medical Journal)* 297(6656): 1070.

———. 1989. Infective Cause of Childhood Leukaemia. *Lancet* 1(8634): 378–79.

———. 1992. Immunosuppressive Therapy and Immunological Disorders. *Cancer Research: American Journal of Cancer, the Official Organ of the American Association for Cancer Research, Inc.* 52(19): 5474s.

———. 1995. Epidemiological Evidence for an Infective Basis in Childhood Leukaemia. *British Journal of Cancer* 71(1): 1–5.

———. 1997a. Infection and Childhood Leukaemia Near Nuclear Sites. *Lancet* 349(9066): 1782.

————. 1997b. High-contact Paternal Occupations, Infection and Childhood Leukaemia: Five Studies of Unusual Population-Mixing of Adults. *British Journal of Cancer* 76(6): 1539–49.

————. 1998. Infection and Childhood Leukemia, Cancer Causes Control 9(3): 237–39.

————. 2000. An Ignoble Lineage: A Review of Mel Greaves, *Cancer: The Evolutionary Legacy*. Oxford: Oxford University Press. *Nature* 405(6783): 122–23.

Kinlen, Leo J., and C. Stiller. 1993. Population Mixing and Excess of Childhood Leukemia. *BMJ (British Medical Journal)* 306(6882): 930.

Kirby, Alex. 1999. Ban DDT Says Wildlife Group: Malaria: DDT Should Be "a Pesticide of Last Resort," Says WWF. BBC World Service online, 27 January.

Klingender, Frances. 1968. Art and the Industrial Revolution. Revised by Arthur Elton. London: Evelyn, Adams & MacKay.

Kmietowicz, Zosia. 2000. Control Malaria to Help Defeat Poverty, Says WHO. *BMJ (British Medical Journal)* 320(7243).

Knox, E. G., and E. A. Gilman. 1998. Migration Patterns of Children with Cancer. *Journal of Epidemiology and Community Health* 52(11): 716–26.

Knutson, Ronald D., C. R. Taylor, J. B. Penson, and E. G. Smith. 1990. *Economic Impacts of Reduced Chemical Use*. College Station: Knutson & Associates.

Kolata, Gina. 1997. Study Discounts DDT Role in Breast Cancer. *New York Times*, 30 October.

————. 1998. The Herbal Potions That Make Science Sick. *New York Times*, 15 November.

————. 1999a. Probing Disease Clusters: Easier to Spot Than Prove, *New York Times*, 31 January.

————. 1999b. Study Inconclusive on Chemicals Effects. *New York Times*, 4 August.

Koning, H. W., K. R. Smith, and J. M. Last. 1985. Biomass Fuel Combustion and Health. *Bulletin of the World Health Organization* 63(1): 11–26.

Koprowski, H., and Michael B. A. Oldstone. 1996a. Prologue. In *Microbe Hunters—Then and Now*, edited by H. Koprowski and Michael B. A. Oldstone. Bloomington: Med-Ed Press.

————. 1996b. Smallpox. In *Microbe Hunters—Then and Now,* edited by H. Koprowski and Michael B. A. Oldstone. Bloomington: Med-Ed Press.

Koshland, Daniel E., Jr. 1989. Scare of the Week. *Science* 244(4900): 9.

Kouwenhoven, John A. 1967. *The Arts in Modern American Civilization*. New York: W.W. Norton.

Kozlowski, Kim. 2000. Professor Defends Genetic Research: Vandals Torched Office of MSU Researcher Working to Boost Output of Nutritious Food. *Detroit News,* 24 January.

Krieger Nancy, Mary S. Wolff, Robert A. Hiatt, Marilyn Rivera, Joseph Vogelman, and Norman Orentreich. 1994. Breast Cancer and Serum Organochlorines: A Prospective Study among White, Black, and Asian Women. *Journal of the National Cancer Institute* 86(8): 589–99.

Ku, Maurice S. B. 2000. Metabolically Modified Rice Exhibits Superior Photosynthesis and Yield. *ISB News Report (Information Systems for Biotechnology)*, May.

Kuiper, George G. J. M., Josephine G. Lemmen, Bo Carlsson, J. Christopher Corton, Stephen H. Safe, Paul T. van der Saag, Bart van der Burg, and Jan-Ake Gustafsson. 1998. Interaction of Estrogenic Chemicals and Phytoestrogens with Estrogen Receptor b. *Endocrinology* 139(10): 4252–62.

Kuiper, Harry A., Hub P. J. M. Noteborn, and A. C. M. Peijnenburg. 1999. Adequacy of Methods for Testing the Safety of Genetically Modified Foods. *Lancet* 354(9187): 1312. .

Lachiusa, M. J. 1999. I Sing of America's Mongrel Culture. *New York Times,* 14 November.

Lachmann, Peter. 1999. Health Risks of Genetically Modified Foods. *Lancet* 354(9172): 69.

Lalumandier, James A., and Leona W. Ayers. 2000. Fluoride and Bacterial Content of Bottled Water vs. Tap Water. *Archives of Family Medicine* 9(3): 246–50.

Lancet. 1997. African Onchocerciasis Eradicated? *Lancet* 350(9092): 1689.

———. 1999a. Autism and Measles, Mumps, and Rubella Vaccine: No Epidemiological Evidence for a Causal Association. *Lancet* 353(9169).

———. 1999b. Food Debate. *Lancet* 354(9191).

———. 2000a. Herbs Not So Harmless. *Lancet* 355(9198).

———. 2000b. St John's Wort—A Double Edged Sword. *Lancet* 355(9203).

———. 2000c. Measles, MMR, and Autism: The Confusion Continues. *Lancet* 355(9213):1379.

———. 2000d. Put That Duster Away! *Lancet* 355(9216).

———. 2000e. Good News in Breast-cancer Mortality. *Lancet* 355(9217).

———. 2000f. Caution Required with the Precautionary Principle. *Lancet* 356(9226): 265.

Law, Susanna. 1995. Risks of Vegan Diets during Pregnancy and Lactation. *AARN News Letter* 51(3): 11.

Layton, Edwin, Jr. 1974. Technology vs. Knowledge. *Technology and Culture* 15(1): 31–41.

Lean, Geoffrey. 2000. Rejected GM Food Dumped on the Poor. *The Independent (London)*, 18 June.

Lehrman, Sally. 1999. As Rockefeller Head Warns of Backlash. *Nature* 401(6756): 831–32.

Lehrman, Sally. 2000. California Targets GM-trial Vandals with New Legislation. *Nature* 404(6780): 799.

Leisinger, Klaus. 1999a. Panelist's Remarks. In *Biotechnology and Biosafety: A Forum*, edited by Ismail Serageldin and Wanda W. Collins. Washington, D.C.: World Bank.

———. 1999b. Disentangling Risk Issues. In *Biotechnology for Developing-Country Agriculture: Problems and Opportunities*, edited by Gabrielle J. Persley. Washington, D.C.: International Food Policy Research Institute.

———. 2000. Can Biotechnology End Hunger—Yes: Stop Blocking Progress. *Foreign Policy* (119): 113–22.

Leopold, A. Carl, and Robert Ardrey. 1972. Toxic Substance in Plants and Food Habits of Early Man. *Science* 176(4034).

Levine, Lawrence W. 1988. *Highbrow/Lowbrow: The Emergence of Cultural Hierarchy in America*. Cambridge: Harvard University Press.

———. 1996. *The Opening of the American Mind: Canons, Culture and History*. Boston: Beacon Press.

Levy, Douglas, and Marin Cipolini. 1997. Antifungal Activity of Solanum Fruit Glycoalkoloids: Implications Frugivory and Seed Dispersal. *Ecology* 78(3): 799–809.

Levy, Stuart. 2000. How Clean Should We Be? International Conference on Emerging Infectious Diseases: ICEID 2000. Atlanta, Ga., 16–19.

Lewis, Martin W. 1996. Radical Environmental Philosophy and the Assault on Reason. In *The Flight from Science and Reason*, edited by Paul R. Gross, Norman Levitt, and Martin W. Lewis. New York: The New York Academy of Sciences.

Liao, Youlian, Daniel L. McGee, Guichan Cao, and Richard S. Cooper. 2000. Quality of the Last Year of Life of Older Adults: 1986 vs 1993. *JAMA (Journal of the American Medical Association)* 283(4): 512–18.

Liburdy, R. P. 1999. Calcium and EMFs: Graphing the Data. *Science* 285(5426): 337.

Lieberman, Adam. 1998. *Facts Versus Fear: A Review of the 20 Greatest Unfounded Health Scares of Recent Times*. Revised edition. New York: American Council on Science and Health.

Lieberman, Philip. 1988. Voice in the Wilderness: How Humans Acquired the Power of Speech. *Sciences* 28(4): 23–29.

Liese, Bernhard. 1995. The Challenge of Success. In *Sustainable Settlement and Development of the Onchocerciasis Control Programme Area: Proceedings of a Ministerial Meeting*, edited by J. Elder and L. Cooley. Technical Paper No. 310. Washington, D.C.: World Bank.

Linden, Martha. 1999. Mowlam Promises Tougher GM Rules. *The Independent (London)*, 18 December.

Linet, M. S., E. E. Hatch, and R. A. Kleinerman. 1997. Residential Exposure to Magnetic Fields and Acute Lymphoblastic Leukemia in Children. *New England Journal of Medicine* 337(1).

Lindsay, James A. 1997. Chronic Sequelae of Foodborne Disease. *Emerging Infectious Diseases Journal* 3(4).

Lipton, Michael. 1999. Genetically Modified Foods Already Reducing Undernutrition. Letter to *Financial Times (London)*, 12 July.

Lockwood, Jeffrey A. 2000. Nontarget Effects of Biological Control: What Are We Trying to Miss? In *Nontarget Effects of Biological Control*, edited by Peter A. Follett and Jian J. Duan. Boston: Kluwer Academic.

Loder, Natasha 1999a. Royal Society: GM Food Hazard Claim Is "Flawed." *Nature* 399(6733): 188.

———— 1999b. Journal under Attack over Controversial Paper on GM Food. *Nature* 401(6755): 731.

Longnecker, M. P., B. C. Gladen, D. G. Patterson, Jr., and W. J. Rogan. 2000. Polychlorinated Biphenyl (PCB) Exposure in Relation to Thyroid Hormone Levels in Neonates. *Epidemiology* 11(3).

Longnecker, M. P., W. J. Rogan, and G. Lucier. 1997. The Human Health Effects of DT (dichlorodiphenyl-trichloroethane) and PCBs (polychlorinated biphenyls) and an Overview of Organochlorines in Public Health. *Annual Review of Public Health* 18: 211–44.

Lopez-Carillo, L., A. Blair, C. M. Lopez, M. Cebrian, C. Rueda, R. Reyes, A. Mohar, and J. Bravo. 1997. Dichlorodiphenyltrichloroethane Serum Levels and Breast Cancer Risk: A Case Control Study from Mexico. *Cancer Research* 57: 3728–32.

Losey, John, Linda S. Rayor, and Maureen E. Carter. 1999. Transgenic Pollen Harms Monarch Larvae. *Nature* 399(6733): 214.

Lovblad, Karl-Olof, Gianpaolo Ramelli, Luca Remonda, Arto C. Nirkko, Christoph Ozdoba, and Gerhard Schroth. 1997. Retardation of Myelination Due to Dietary Vitamin B12 Deficiency: Cranial MRI Findings. *Pediatric Radiology* 27(2): 155–58.

Lowenthal, David. 1979. Age and Artifact: Dilemma of Appreciation. In *The Interpretation of Ordinary Landscapes: Geographical Essays*, edited by D. W. Meinig. New York: Oxford University Press.

————. 1985. *The Past Is a Foreign Country*. Cambridge: Cambridge University Press.

Lu, Frank C. 1999. Irradiated Food for the Protection of Human Health with Special Emphasis on Irradiated Meat. *Technology: Journal of Science Serving Legislative, Regulatory, and Judicial Systems* 6(2-3): 89–94.

Lukacs, John L. 1993. *The End of the Twentieth Century and the End of the Modern Age*. New York: Ticknor & Fields.

Lucky, Robert. 2000. The Quickening of Science Communication *Science* 289(5477): 259–64.

Lutchman, L., V. Inyang, and D. Hodgkinson. 1999. Phytophotodermatitis Associated with Parsnip Picking. *Journal of Accident and Emergency Medicine* 16(6): 453–54.

Lutter, Randall. 1999. Food Irradiation—The Neglected Solution to Food-Borne Illness. *Science* 286(5449): 275–76.

MacIlwain, Colin. 1999a. Developing Countries Look for Guidance in GM Crops Debate. *Nature* 401(6756): 831–32.

———. 1999b. Access Issues May Determine Whether Agri-biotech Will Help the World's Poor. *Nature* 402(6760).

MacKenzie, Debora. 1999. Friend or Foe, "Friendly" Bacteria Make Terrific Insecticides. But Just a Few Genes Separate the Good Guys from the Bad Guys. *New Scientist* 164(2207).

MacKerron, D. K. L., J. M. Duncan, J. R. Hillman, G. R. Mackay, D. J. Robinson, D. L. Trudgill, and R. J. Wheatley. 2000. Organic Farming: Science and Belief. SCRI (Scottish Crop Research Institute) 1998/99 Annual Report 60-72, Invergowie, Dundee, Scotland, January.

Maddox, John Royden. 1998. *What Remains to be Discovered: Mapping the Secrets of the Universe, the Origins of Life, and the Future of the Human Race.* New York: Martin Kessler Books.

Madigan, Charles M. 1999. "Optimism Gap" Still Strong As 1999 Begins: Polls Indicate All Is Well in Americans' Eyes. *Chicago Tribune* News Service, *Houston Chronicle,* 2 January.

Malaspina, Alex. 1992. Foreword. In *Safety of Water Disinfection: Balancing Chemical and Microbial Risks,* edited by Gunther F. Cruan. Washington, D.C.: ILSI (International Life Sciences Institute) Press.

Malcom, Alan D. B. 1999. Health Risks of Genetically Modified Foods. *Lancet* 354(9172): 69.

Marglin, F. A. 1990. Smallpox in Two Systems of Knowledge. In *Dominating Knowledge: Development, Culture, and Resistance,* edited by F. A. Marglin and S. A. Marglin. Oxford: Clarendon Press.

Marglin, S. A. 1990. Towards the Decolonization of the Mind. In *Dominating Knowledge: Development, Culture, and Resistance,* edited by F. A. Marglin and S. A. Marglin. Oxford: Clarendon Press.

———. 1996. Farmers, Seedsmen, and Scientists: Systems of Agriculture and Systems of Knowledge. In *Decolonizing Knowledge: From Development to Dialogue,* edited by F. A. Marglin and S. A. Marglin. Oxford: Clarendon Press.

Margulis, Lynn, and Dorian Sagan. 1986. *Micro-cosmos: Four Billion Years of Microbial Evolution.* New York: Summit Books.

Mark, Robert, and David P. Billington. 1989. Structural Imperative and the Origin of New Form. *Technology and Culture* 30(2): 300–329.

Marshack, Alexander. 1972. *Roots of Civilization: The Cognitive Beginnings of Man's First Art, Symbol and Notation.* New York: McGraw-Hill.

Martens, Pim, and Lisbeth Hall. 2000. Malaria on the Move: Human Population Movement and Malaria Transmission. *Emerging Infectious Diseases Journal* 6(2).

Marwick, Charles. 1997. Putting Money Where the US Mouth Is: Initiative on Food Safety Gets Under Way. *JAMA (Journal of the American Medical Association)* 277(17): 1340–42.

———. 2000a. Genetically Modified Crops Feed Ongoing Controversy. *JAMA (Journal of the American Medical Association)* 283(2).

———. 2000b. International Plan Focuses on Eradication of Polio and Containment of the Virus. *JAMA (Journal of the American Medical Association)* 283(12).

Masood, Ehsan. 1999a. Gag on Food Scientist Is Lifted as Gene Modification Row Hots Up. *Nature* 397(6720): 547.

———. 1999b. Food Scientist in GMO Row Defends "Premature" Warning. *Nature* 398(6723): 98.

Matossian, Mary K. 1989. *Poisons of the Past: Molds, Epidemics and History.* New Haven: Yale University Press.

Matricardi, P. M., F. Rosmini, S. Riondino, M. Fortini, L. Ferrigno, M. Rapicetta, and S. Bonini. 2000. Exposure to Foodborne and Orofecal Microbes Versus Airborne Viruses in Relation to Atopy and Allergic Asthma: Epidemiological Study. *BMJ (British Medical Journal)* 320(7232): 412–17.

May, Sir Robert. 2000. Epidemiology Enhanced: Simple Rules with Complex Dynamics. *Science* 287(5453): 601–2.

Mayeno, Arthur N., and Gerald J. Gleich. 1994. Eosinophilia-myalgia Syndrome and Tryptophan Production: A Cautionary Tale. *TIBTECH (Trends in Biotechnology).* 12(9): 346–52.

Mays, Patricia J. 1999. The Workplace Has Become Safer in the 20th Century, CDC says. Associated Press, Nando Media online, 11 June.

Maysich, K. B., C. B. Ambrosone, J. E. Vena, P. G. Shields, P. Mendola, P. Kostyniak, H. Greizerstein, S. Graham, J. R. Marshall, E. F. Schisterman, and J. L. Freudenheim. 1998. Environmental Organochlorine Exposure and Post-menopausal Breast Cancer Risk. *Cancer Epidemiology Biomarkers & Prevention* 7: 181–88.

Mazur, Barbara, Enno Krebbers, and Scott Tingey. 1999. Gene Discovery and Product Development for Grain Quality Traits. *Science* 285(5426): 372–75.

McCann, Kevin. 2000. Review Article: The Diversity-stability Debate. *Nature* 405(6783): 228–33.

McClellan, James E., and Harold Dorn. 1999. *Science and Technology in World History: An Introduction.* Baltimore: Johns Hopkins University Press.

McEvoy, P. B., and E. M. Coombs. 2000. Why Things Bite Back: Unintended Consequences of Biological Control. In *Nontarget Effects of Biological Control,* edited by Peter A. Follett and Jian J. Duan. Boston: Kluwer Academic.

McFalls, Joseph A. 1998. Population: A Lively Introduction, 3d ed. *Population Bulletin* 53(3): 1–48.

McGinnis, J. M., and W. H. Foege. 1993. Actual Causes of Death in the United States. *JAMA: Journal of the American Medical Association* 270(18): 2207–11.

McGrew, William C. 1993. The Intelligent Use of Tools: Twenty Propositions. In *Tools, Language, and Cognition in Human Evolution*, edited by Kathleen R. Gibson and Tim Ingold. New York: Cambridge University Press.

McLachlan, John A. 1997. Synergistic Effects of Environmental Estrogen: Report Withdrawn. Letter. *Science* 277(5325): 462–64.

Mead P. S., L. Slutsker, V. Dietz, L. F. McCaig, J. S. Bresee, C. Shapiro, P. M. Griffin, and R. V. Tauxe. 1999. Food-related Illness and Death in the United States. *Emerging Infectious Diseases Journal* 5(5): 607–25.

Melchart, Dieter, Ellen Walther, Klaus Linde, Roland Brandmaier, and Christian Lersch. 1998. Echinacea Root Extracts for the Prevention of Upper Respiratory Tract Infections: A Double-blind, Placebo-Controlled Randomized Trial. *Archives of Family Medicine* 7(7): 541–45.

Mellanby, Kenneth. 1992. *The DDT Story*. Farnham, Surrey and Bracknell, Berkshire: British Crop Protection Council.

Mendoza, Martha 1997. Toughest Child Labor Laws Are Not Enforced. Associated Press, Nando.net online, 19 December.

Metcalf, R. L. 1980. Changing Role of Insecticides in Crop Production. *Annual Review of Entomology* 25: 219–56.

———. 1986. The Ecology of Insecticides and the Chemical Control of Insects. In *Ecological Theory and Integrated Pest Management Practice*, edited by Marcos Kogan. New York: John Wiley.

Miller, Henry I. 1997. When Worlds Collide: Science, Politics, and Biotechnology. *Priorities: For Long Life and Good Health* 9(4): 8–13.

———. 1999. GM crops: The Unexpected Arm of the Bio-police. *Financial Times (London)*, 21 December.

———. 2000a. Anti-biotech Sentiment Has Its Own Risks. *Financial Times (London)*, 22 March.

———. 2000b. NAS Report under Scrutiny. *Science* 288(5470): 1343.

Miller, Henry I., and Gregory Conko. 2000. The Protocol's Illusionary Principle. *Nature Biotechnology* 18(4): 360–69.

Miller, M., and G. Applet. 1993. Biological Control: A Little Knowledge Is a Dangerous Thing. *Rutgers Law Review* 45: 285.

Milton, Katharine. 1981. Distribution Pattern of Tropical Food Plants as an Evolutionary Stimulus to Primate Mental Development. *American Anthropologist* 83(3)534–48.

———. 1987. Primate Diets and Gut Morphology: Implications for Hominid Evolution. In *Food and Evolution: Toward a Theory of Human Food*

Habits, edited by Marvin Harris and Eric B. Ross. Philadelphia: Temple University Press.

———. 1988. Foraging Behavior and the Evolution of Primate Intelligence. In *Machiavellian Intelligence: Social Expertise and the Evolution of Intellect in Monkeys, Apes, and Humans*, edited by Richard Byrne and Andrew Whitten. Oxford: Clarendon Press.

———. 1993. Diet and Primate Evolution. *Scientific American* 269(2): 86–93.

———. 1999. A Hypothesis to Explain the Role of Meat-eating in Human Evolution. *Evolutionary Anthropology* 8(1): 11.

Milton, Katharine, and M. W. Demment. 1988. Digestion and Passage Kinetics of Chimpanzees Fed High and Low-fiber Diets and Comparison with Human Data. *Journal of Nutrition* 118(9): 1082–88.

Mitchell, Donald O., Merlinda D. Ingco, and Robert C. Duncan. 1998. *The World Food Outlook*. Cambridge: Cambridge University Press.

Mitchell, Wesley C. 1941. Conservation, Liberty and Economics. In *National Wildlife Federation, Foundations of Conservation Education*. NWF Committee on Conservation Education Pamphlet No. 3. Washington: National Wildlife Federation.

Mithen, Steven J. 1996. *The Prehistory of the Mind: The Cognitive Origins of Art, Religion and Science*. London: Thames & Hudson.

MMWR. 1997. Outbreaks of *Escherichia coli* O157: H7 Infection and Cryptosporidiosis Associated with Drinking Unpasteurized Apple Cider-Connecticut and New York, October 1996. *Morbidity and Mortality Weekly Report* 46(1): 4–8.

———. 1999a. Ten Great Public Health Achievements in the 20th Century. *Morbidity and Mortality Weekly Report* 48.

———. 1999b. Achievements in Public Health, 1900–1999: Impact of Vaccines Universally Recommended for Children-United States, 1990–1998. *Morbidity and Mortality Weekly Report* 48(12): 243–48.

———. 1999c. Achievements in Public Health, 1900–1999: Motor-vehicle Safety: A 20th Century Public Health Achievement. *Morbidity and Mortality Weekly Report* 48(18): 369–74.

———. 1999d. Achievements in Public Health, 1900–1999: Improvements in Workplace Safety-United States, 1900–1999. *Morbidity and Mortality Weekly Report* 48(22): 461–69.

———. 1999e. Achievements in Public Health, 1900–1999: Control of Infectious Diseases. *Morbidity and Mortality Weekly Report* 48(29): 621–29.

———. 1999f. Achievements in Public Health, 1900–1999: Decline in Deaths from Heart Disease and Stroke-United States, 1900–1999. *Morbidity and Mortality Weekly Report* 48(30): 649.

———. 1999g. Achievements in Public Health, 1900–1999: Healthier Mothers and Babies. *Morbidity and Mortality Weekly Report* 48(38): 849–58.

———. 1999h. Achievements in Public Health, 1900–1999: Safer and Healthier Foods. *Morbidity and Mortality Weekly Report* 48(40): 905–13.

———. 1999i. Achievements in Public Health, 1900–1999: Fluoridation of Drinking Water to Prevent Dental Caries. *Morbidity and Mortality Weekly Report* 48(41): 933–40.

———. 1999j. Achievements in Public Health, 1900–1999: Tobacco Use - United States, 1900–1999. *Morbidity and Mortality Weekly Report* 48(43): 986–93.

———. 1999k. Achievements in Public Health, 1900–1999: Family Planning. *Morbidity and Mortality Weekly Report* 48(47): 1073–80.

———. 1999l. Achievements in Public Health, 1900–1999: Changes in the Public Health System. *Morbidity and Mortality Weekly Report* 48(50): 1141–47.

———. 1999m. Outbreak of Salmonella Serotype Muenchen Infections Associated with Unpasteurized Orange Juice: United States and Canada, June 1999. *Morbidity and Mortality Weekly Report* 48(27): 582–85.

———. 2000a. Developing and Expanding Contributions of the Global Laboratory Network for Poliomyelitis Eradication, 1997–1999. *Morbidity and Mortality Weekly Report* 49(8): 156–60.

———. 2000b. Preliminary FoodNet Data on the Incidence of Foodborne Illnesses: Selected Sites, United States, 1999. *Morbidity and Mortality Weekly Report* 49(10): 201.

Moffat, Anne Simon. 1999a. Engineering Plants to Cope with Metals. *Science* 285(5426): 369–70.

———. 1999b. Crop Engineering Goes South, *Science* 285(5426): 370–71.

Mokyr, Joel. 1994. Progress and Inertia in Technological Change. In *Capitalism in Context: Essays on Economic Development and Cultural Change in Honor of R. M. Hartwell,* edited by John A. James and Mark Thomas. Chicago: University of Chicago Press.

———. 1998. Innovation and Its Enemies: The Economic and Political Roots of Technological Inertia. In *The Not-so Dismal Science,* edited by Mancur Olson and Satu Kakkaren. New York: Oxford University Press.

Molyneux, D. H. 1995. Onchocerciasis Control in West Africa: Current Status and Future of the Onchocerciasis Control Programme. *Parasitology Today* 11(11): 399–402.

Molyneux, D. H., and J. B Davies. 1997. Onchocerciasis Control: Moving towards the Millennium. *Parasitology Today* 13(11): 418–25.

Molyneux, D. H., and C. Morel. 1998. Onchocerciasis and Chagas' Disease Control: The Evolution of Control Via Applied Research Through Changing Development Scenarios. *British Medical Bulletin* 54(2): 327–39.

Monbiot, George. 1999. Whose Wildlife Is It Anyway. *Electronic Mail & Guardian.* 9 June.

————. 2000. Beware the Appliance of Science: These Scientists Act As If They Live on Another Planet to the Rest of Us. *The Guardian (London),* 24 February.

Moore, Barrington, Jr. 1978. *Injustice: The Social Basis of Obedience and Revolt.* White Plains: M. E. Sharpe.

Moore, Stephen, and Julian L. Simon. 1999. The Greatest Century That Ever Was: 25 Miraculous Trends of the Past 100 Years. *Policy Analysis* 364: 1–32.

Morrison, W. Ivan. 1999. Biotechnology and Animal Vaccines. In *Biotechnology for Developing-Country Agriculture: Problems and Opportunities,* edited by Gabrielle J. Persley. Washington, D.C.: International Food Policy Research Institute.

Munkvold, Gary P., and Richard L. Hellmich. 1999. Genetically Modified, Insect Resistant Corn: Implications for Disease Management. APSnet Feature (American Phytopathological Society), 15 October.

Murray, Christopher J. L., and Alan D. Lopez. 1996b. Evidence-based Health Policy: Lessons from the Global Burden of Disease Study. *Science* 274(5288): 740–43.

Murray, Christopher J. L., and Alan D. Lopez, eds. 1996a. The Global Burden of Disease: A Comprehensive Assessment of Mortality and Disability from Diseases, Injuries, and Risk Factors in 1990 and Projected to 2020. Cambridge: The Harvard School of Public Health on behalf of the World Health Organization and the World Bank.

Nabakwe, Ruth. 1998.Treaty on Hazardous Chemicals Imposes Trade Controls. *PANA,* Panafrican News Agency online, 14 September.

Nandy, Ashis, and Shiv Visvanathan. 1990. Modern and Its Non-modern Critics: A Study in Discourse. In *Decolonizing Knowledge: From Development to Dialogue,* edited by F. A. Marglin and S. A. Marglin. Oxford: Clarendon Press.

NAS. 1971. *The Life Sciences: Recent Progress and Application to Human Affairs: The World of Biological Affairs: Requirements for the Future.* Washington, D.C.: National Academy of Sciences, Committee on Research in Life Sciences of the Committee on Science and Public Policy.

————. 1973. *Toxicants Occurring Naturally in Foods.* Washington, D.C.: National Academy of Sciences, Committee on Food Protection, Food and Nutrition Board, National Research Council.

————. 1988. *Quality Protein Maize.* Washington, D.C.: National Academy of Sciences, Board on Science and Technology for International Development.

————. 1989. Field Testing Genetically Modified Organisms: Framework for Decisions. Committee on Scientific Evaluation of the Introduction of Genetically Modified Microorganisms and Plants into the Environment, Board on Biology, Commission on Life Sciences, National Research Council. Washington, D.C.: National Academy Press.

————. 1999. Effects of Hormonally Active Agents in the Environment Uncertain, Press release: National Research Council, Commission On Life Sciences, Board on Environmental Studies and Toxicology, Committee on Hormonally Active Agents in the Environment, 3 August.

————. 2000a. Press Release: Study Finds Strong Evidence That Exposure to Some Indoor Substances Can Lead to or Worsen Asthma. Washington, D.C.: National Academy of Sciences.

————. 2000b. Press Release: Genetically Modified Pest-protected Plants: Science and Regulation. Washington, D.C.: National Research Council, National Academy of Sciences.

————. 2000c. Press release: Antioxidants' Role in Preventing Chronic Disease Still Uncertain. Washington, D.C.: National Academy of Sciences, Institute of Medicine.

National Wildlife Federation. 1941. *The Foundations of Conservation Education*. NWF Committee on Conservative Education, Pamphlet No. 3. Washington, D.C.: National Wildlife Federation.

NB. 1999. A Golden Bowl of Rice. *Nature Biotechnology* 17(9): 831.

————. 2000. Swapping Science for Consensus in Montreal. *Nature Biotechnology* 18(3): 239.

NCI. 2000. Annual Report Shows Continuing Decline in U.S. Cancer Incidence and Death Rates. Washington, D.C.: National Cancer Institute, National Institutes of Health.

Nduru, Moyiga. 1999. Malaria Epidemic Rife in Southern Africa, MISAnet/Inter Press Service. *Woza.com* online, 16 March.

Nelson, Buck R. 1977. Bodie: A Ghost Town Stabilized. In *Ghost Towns and Mining Camps: Selected Papers*. Washington, D.C.: Preservation Press for the National Trust for Historic Preservation.

Newcomb, W. W., Jr. 1986. *The Indians of Texas: From Prehistoric to Modern Times*. Austin: University of Texas Press.

NIAID. 2000. The Jordan 2000 Report: Accelerated Development of Vaccines. Division of Microbiology and Infectious Diseases, National Institute of Allergy and Infectious Diseases, National Institutes of Health, Thursday, 2 March.

NIEHS. 1997. Asthma and Its Environmental Triggers: Scientists Take a Practical New Look at a Familiar Illness, NIEHS Fact Sheet #9, National Institute of Environmental Health Sciences, National Institutes of Health, July.

————. 1999. Environmental Health Institute Report Concludes Evidence Is "Weak" That Electric and Magnetic Fields Cause Cancer, National Institute of Environmental Health Sciences, Press release, 15 June.

Nolan, Christopher. 1988. *Under the Eye of the Clock*. New York: St. Martin's Press.

Nordhaus, William D. 1997. Do Real-output and Real-wage Measures Capture Reality?: The History of Lighting Suggests Not. In *The Economics of New Goods*, edited by Timothy J. and Robert J. Gordon, 29–66. Chicago: University of Chicago Press for the National Bureau of Economic Research, Conference on Research in Income and Wealth, Studies in Income and Wealth, 58.

Normile, Dennis. 1999. Rice Biotechnology: Rockefeller to End Network after 15 Years of Success. *Science* 286(5444): 1468–69.

———. 2000. Hopes Grow for Hybrid Rice to Feed Developing World. *Science* 288(5465): 429.

North, Geoffrey. 1990. The Race for Resistance Genes. *Nature* 347(6293): 517.

North, K., and J. Golding. 2000. A Maternal Vegetarian Diet in Pregnancy Is Associated with Hypospadias. *BJU International (British Journal of Urology)* 85(1): 107–13.

Nortier, J. L., M.-C. Muniz Martinez, H. H. Schmeiser, V. M. Arlt, C. A. Bieler, M. Petein, M. F. Depierreux, L. De Pauw, D. Abramowicz, P. Vereerstraeten, and J.-L. Vanherweghem. 2000. Urothelial Carcinoma Associated with the Use of a Chinese Herb (Aristolochia Fangchi). *New England Journal of Medicine* 342(23).

NR. 1990. Inadequate Vegan Diets at Weaning. *Nutrition Review* 48(8): 323–26.

NRC. 1989. Alternative Agriculture. Board on Agriculture, Committee on the Role of Alternative Farming, Methods in Modern Agriculture. Washington, D.C.: National Academy Press.

———. 1996a. Carcinogens and Anticarcinogens in the Human Diet: A Comparison of Naturally Occurring and Synthetic Substances. Committee on Comparative Toxicity of Naturally Occurring Carcinogens, Board on Environmental Studies and Toxicology, and the Commission on Life Sciences, National Research Council. Washington, D.C.: National Academy Press.

———. 1996b. Possible Health Effects of Exposure to Residential Electric and Magnetic Fields, Executive Summary. Washington, D.C.: National Research Council, National Academy Press.

———. 1997. Possible Health Effects of Exposure to Residential Electric and Magnetic Fields. Washington, D.C.: National Research Council, National Academy of Sciences Press.

———. 1999a. Hormonally Active Agents in the Environment. Washington, D.C.: National Research Council, Committee on Hormonally Active Agents in the Environment, National Academy of Sciences Press.

———. 1999b. Genetically Modified Pest Protected Plants. Washington, D.C.: National Research Council, Board on Agriculture and Natural Resources, National Academy of Sciences Press.

Nwoke, B. E. B. 1990. Socio-economic Aspects of Human Onchocerciasis in Africa: Present Appraisal. *Journal of Hygiene, Epidemiology, Microbiology, and Immunology* 34(1): 37–44.

Oaks, Stanley, Jr., Violaine S. Mitchell, Greg W. Pearson, and Charles C. J. Carpenter, eds. 1991. *Malaria: Obstacles and Opportunities*. Washington, D.C.: National Academy Press.

Oates, Joyce Carol. 1973. New Heaven and Earth. *Arts in Society* special issue, The Humanist Alternative 10(1): 36–43, reprinted from *Saturday Review*/The Arts 55(45): 51–54, 4 November 1972.

OECD. 1999. The Concept of Substantial Equivalence in the Safety Assessment of Novel Foods. OECD, Media Relations Division Organization for Economic Co-operation and Development, 6 October.

———. 2000. Report of the Task Force for the Safety of Novel Food and Feeds. Organization for Economic Co-operation and Development, Task Force for the Safety of Novel Food and Feeds, 17 May.

Oerke, E. C., H. W. Dehne, F. Schohnbeck, and A. Weber. 1994. *Crop Production and Crop Protection: Estimated Losses in Major Food and Cash Crops*. Amsterdam: Elsevier.

O'Hara, Mary Ann, David Kiefer, Kim Farrell, and Kathi Kemper. 1998. A Review of 12 Commonly Used Medicinal Herbs. *Archives of Family Medicine* 7(7): 523–36.

Oldstone, Michael B. A. 1998. *Viruses, Plagues, and History*. New York: Oxford University Press.

Ong, Walter J. 1982. Orality and Literacy: The Technologizing of the Word. London: Methuen.

Osterholm, M. T., and M. E. Potter. 1997. Irradiation Pasteurization of Solid Foods: Taking Food Safety to the Next Level. *Emerging Infectious Diseases Journal* 3(4): 23–50.

Paehlke, Robert C. 1989. *Environmentalism and the Future of Progressive Politics*. New Haven: Yale University Press.

PAFL. 1999. Vatican Calls for Honesty on Biotechnology: "Don't Fear Scientific Progress." Rome: Pontifical Academy for Life, October 12.

Palevitz, Barry A. 1999. Bt or not Bt: Transgenic Corn vs. Monarch Butterflies: Could a Diet Rich in Transgenic Corn Pollen Doom the Monarch butterfly? *Scientist* 13(12).

PANA. 2000. Malaria Kills More Than African Conflicts or Disasters. PANA (Panafrican News Agency), 30 June.

Papp, Leslie. 2000. Coroners to Track Alternative Medicine Cases: Ontario Responds to Star Series on Unorthodox Cures. *Toronto Star*, 22 April.

Park, Robert L. 2000a. *Voodoo Science*. New York: Oxford University Press.

———. 2000b. Voodoo Science and the Power-line Panic. *Forbes Magazine* 165(12).

Patterson, Orlando. 1994a. *Global Culture and the American Cosmos*. New York: Andy Warhol Foundation for the Visual Arts, Paper Series on the Arts, Culture, and Society, 2.

————. 1994b. Ecumenical America: Global Culture and the American Cosmos. *World Policy Journal* 11(3).

Paunio, M., O. P. Heinonen, M. Virtanen, P. Leinikki, A. Patja, and H. Peltola. 2000. Measles History and Atopic Diseases: A Population-based Cross-sectional Study. *JAMA (Journal of the American Medical Association)*. 283(3).

Peirce, Charles S. 1984. Some Consequences of Four Incapacities. *Journal of Speculative Philosophy* 2: 140–57, 1868; Reprinted in *Writings of Charles S. Peirce: A Chronological Edition*, edited by Max H. Fisch 2: 211–42, 1867–1871. Bloomington: Indiana University Press.

————. 1992. How to Make Our Ideas Clear. *Popular Science Monthly* 12(1): 286–302, January 1878, reprinted in *The Essential Peirce: Selected Philosophical Writings*, edited by Nathan Houser and Christian J. W. Kloesel. Bloomington: Indiana University Press.

Peng, J., D. E. Richards, N. M. Hartley, G. P. Murphy, K. M. Devos, J. E. Flintham, J. Beales, L. J. Fish, A. J. Worland, F. Pelica, D. Sudhakar, P. Christou, J. W. Snape, M. D. Gale, and N. P. Harberd. 1999. "Green revolution" Genes Encode Mutant Gibberellin Response Modulators. *Nature* 400(6741): 256–61.

Perfecto, Ivette. 1992. Farm Workers, Pesticides and the International Connection. In *Race and the Incidence of Environmental Hazards: A Time for Discourse*, edited by Bunyan Bryant and Paul Mohai. Boulder: Westview Press.

Peries, Herve, and Sandy Cairncross. 1997. Global Eradication of Guinea Worm. *Parasitology Today* 13(11): 431–37.

Persley, Gabrielle J. 2000. Agricultural Biotechnology and the Poor: Promethean Science. In *Agricultural Biotechnology and the Poor*, edited by Gabrielle J. Persley and M. M. Lantin. Washington, D.C.: Consultitative Group on International Agricultural Research and the U.S. National Academy of Sciences.

Peto, R., J. Boreham, M. Clarke, C. Davies, and V. Beral. 2000a. UK and USA Breast Cancer Deaths Down 25% in Year 2000 at ages 20–69 years. *Lancet* 355(9217): 1822–30.

————. 2000b. Favourable and Unfavourable Effects on Long-term Survival of Radiotherapy for Early Breast Cancer: An Overview of the Randomised Trials. *Lancet*. 355(9217): 1757–70.

Petroski, Henry. 1985. *To Engineer Is Human: The Role of Failure in Successful Design*. New York: St. Martin's Press.

Pimentel, David S., and Peter H. Raven. 2000. Commentary: Bt corn Pollen Impacts on Nontarget Lepidoptera: Assessment of Effects in Nature. *PNAS (Proceedings of the National Academy of Sciences)* 97(15): 8198–99.

Pinker, Steven. 1994. *The Language Instinct*. New York: William Morrow and Company Inc.

————. 1997. *How the Mind Works*. New York: W.W. Norton.

Pinstrup-Andersen, Per. 1999a. Modern Biotechnology and Small Farmers in Developing Countries. *International Food Policy Research Institute (IFPRI) Research Perspectives* 21(2).

————. 1999b. Agricultural Biotechnology, Trade, and the Developing Countries. *AgBioForum* 2(3&4): 215–17.

Pinstrup-Andersen, Per, and Marc J. Cohen. 2000. Modern Bio-Technology for Food and Agriculture: Risks and Opportunities for the Poor. In *Agricultural Biotechnology and the Poor*, edited by Gabrielle J. Persley and M. M. Lanti. Washington, D.C.: Consultitative Group on International Agricultural Research and the US National Academy of Sciences.

Piscitelli, S. C., A. H. Burstein, D. Chaitt, R. M. Alfaro, and J. Falloon. 2000. Indinavir Concentrations and St. John's Wort. *Lancet* 355(9203): 547.

Plato. 1952. *Gorgias*. Trans. W. C. Helmbold. New York: Liberal Arts Press.

————. 1956. *Phaedras*. Trans. W. C. Helmbold and W. G. Rabinowitz. Indianapolis: Bobbs-Merrill Co.

————. 1960. *The Laws*. Trans. A. E. Taylor. New York: E.P. Dutton & Co.

Pois, Robert A. 1986. *National Socialism and the Religion of Nature*. New York: St. Martin's Press.

Pollack, Andrew. 1999. A Disputed Study Suggests Possible Harm: From Genetically Altered Food. *New York Times*, 15 October.

Pollan, Michael. 1994. Against Nativism: Horticultural Formalism May Be Out, But the New American Garden Free of Foreign Flora and Human Artifice, Isn't as Natural as Its Advocates Claim. *New York Times Magazine,* 52–55, 15 May.

Populi. 1998. WHO Predicts, Healthier, Longer Lives in New Millennium. Populi: *The UNFPA Magazine* 25(2): 4–5.

Postman, Neil. 1992. *Technopoly: The Surrender of Culture to Technology.* New York: Knopf.

Postrel, V. I. 1998. *The Future and Its Enemies: The Growing Conflict over Creativity, Enterprise, and Progress.* New York: Free Press.

Potrykus, Ingo, Paola Lucca, Xudong Ye, Salim Al-Babili, Richard F. Hurrel, and Peter Beyer. 1999. Contributions to Food Security by Genetic Engineering with Rice. 16 International Botanical Congress, Saint Louis, Missouri, 1–7 August.

Prakash, C. S. 1999a. Foes of Biotechnology Ignore Global Hunger. *Atlanta Journal-Constitution*, 5 December.

————. 1999b. Feeding a World of Six Billion. *AgBioForum* 2(3&4): 223–25.

PRB. 1997. Chronic Disability Declines Dramatically among U.S. Elderly. *Population Today: News, Numbers, and Analysis* (Population Reference Bureau) 25(9).

Preston, Samuel H., and Michael R. Haines. 1991. *Fatal Years: Child Mortality in Late Nineteenth Century American.* Princeton: Princeton University Press.

Priorities. 1996. Come Back, Meryl Streep, *Priorities: For Long Life and Good Health* 8(4): 6–7.

Proctor, Robert. 1988. *Racial Hygiene: Medicine under the Nazis.* Cambridge: Harvard University Press.

———. 1995. *Cancer Wars: How Politics Shapes What We Know and Don't Know about Cancer.* New York: Basic Books.

———. 1999. *The Nazi War on Cancer.* Princeton: Princeton University Press.

Promislow, Daniel E. L. 1998. Longevity and the Barren Aristocrat. *Nature* 396(6713): 719–20.

Pud'Hom, J.-M., P. Enyong, M. Boussinesq, J.-P. Chippaux, and J. Prod'Hom. 1993. Entomological Evaluation of Ivermectin Mass Treatment against Onchocerciasis. *Medical and Veterinary Entomology* 7(4): 387–89.

Putnam, Susan W., and Jonathan Baert Wiener. 1997. Seeking Safe Drinking Water. In *Risk vs. Risk: Tradeoffs in Protecting Health and the Environment*, edited by John D. Graham and Jonathan Baert Wiener. Cambridge: Harvard University Press.

Quddus, Munir 1999. Child Labor and Global Business: Lessons from the Apparel Industry in Bangladesh. *Journal of Asian Business* 15(4): 81–90.

Radaway, Janice. 1992. Mail-order Culture and Its Critics: The Book-of-the-Month Club, Commodification and Consumption, and the Problem of Cultural Authority. In *Cultural Studies*, edited by Lawrence Grossberg, Cary Nelson, and Paula Triechler. New York: Routledge.

Rahman, Mohammad M., Rasheda Khanam, and Nur Uddin Absar 1999. Child Labor in Bangladesh: A Critical Appraisal of Harkin's Bill and the MOU-type Schooling Program. *Journal of Economic Issues* 33(4): 985–1003.

Rasula, Jed. 1990. Nietzsche in the Nursery: Naïve Classics and Surrogate Parents in Postwar American Cultural Debates. *Representations* 29 (winter): 50–77.

Rauschning, Herman 1940. *Gesprache Mit Hitler.* New York: Europa Verlag.

Ray, Prafulla Chandra. 1902–3 & 1909. *History of Hindu Chemistry.* Calcutta: The Bengal Chemical and Pharmaceutical Works Limited, 1 and 2.

Reader, John. 1998. *Africa: A Biography of the Continent.* New York: Alfred Knopf.

REB. 1996. Onchocerciasis: Progress Towards Elimination in the Americas. *Releve Epidemiologique Hebdomadaire* 71(37).

Reddy, K. S. 1999. The Burden of Disease among the Global Poor. *Lancet* 354(9188).

Reeves, Richard. 1997. Wheels and the Man: A Moving Melodrama, 16 October.

Reisch, Marc. 1999. I Lost It at the Movies. *Chemical & Engineering News* 77(5): 20.

Reiser, Stanley J. 1978. *Medicine and the Reign of Technology*. New York: Cambridge University Press.

————. 1984. The Machine at the Bedside: Technological Transformations of Practices and Values. In *The Machine at the Bedside: Strategies for Using Technology in Patient Care*, edited by Stanley Reiser and Michael Anbar. New York: Cambridge University Press.

Reiter, Paul. 2000. From Shakespeare to Defoe: Malaria in England in the Little Ice Age. *Emerging Infectious Diseases Journal* 6(1).

Remme, J. H. F. 1995. The African Programme for Onchocerciasis Control: Preparing to Launch. *Parasitology Today* 11(11): 403.

Repacholi, Michael H., and Anders Ahlbom. 1999. Link between Electromagnetic Fields and Childhood Cancer Unresolved. *Lancet* 354(9194).

RFSTE. 2000. U.S. Government Dumping Genetically Engineered Corn-soya Mix on Victims of Orissa Super-cyclone. New Delhi: Press release, Diverse Women for Diversity, The Research Foundation for Science, Technology and Ecology (www.vshiva.org) 2 June.

RH. 2000. Vegetarian Diet in Pregnancy Linked to Birth Defect. Reuters Health online, 3 February.

Rhodes, James M. 1980. *The Hitler Movement: A Modern Millenarium Revolution*. Stanford: Hoover Institution Press.

Rhodes, Martha E. 1979. The "Natural" Food Myth. *Sciences* 19(5).

Rice, Eugene W., Robert M. Clark, and Clifford H. Johnson. 1999. Chlorine Inactivation of *Escherichia coli* O157:H7. *Emerging Infectious Diseases Journal* 5(3).

Richer, David L. 2000. Intellectual Property Protection: Who Needs It. In *Agricultural Biotechnology and the Poor*, edited by Gabrielle J. Persley and M. M. Lantin. Washington, D.C.: Consultitative Group on International Agricultural Research and the U. S. National Academy of Sciences.

RICPQL. 1996. *Caring for the Future: Making the Next Decades Provide a Life Worth Living: Report of the Independent Commission on Population and the Quality of Life* . New York: Oxford University Press.

Ries, L. A. G., P. A. Wingo, D. S. Miller, H. L. Howe, H. K. Weir, H. M. Rosenberg, S. W. Vernon, K. Cronin, and B. K. Edwards. 2000. Annual Report to the Nation on the Status of Cancer, 1973–1997. *Cancer: A Journal of the American Cancer Society* 88(10): 2398–424.

Riley, James C. 1989. Sickness, Recovery and Death: A History and Forecast of Ill Health. Iowa City: University of Iowa Press.

Ritter, Leo. 1997. Report of a Panel on the Relationship between Public Exposure to Pesticides and Cancer. *Cancer (American Cancer Society)* 80(10): 2019–33.

Roberts, D. R. , L. L. Laughlin, P. Hsheih, and L. J. Legters. 1997. DDT, Global Strategies, and a Malaria Control Crisis in South America. *Emerging Infectious Diseases Journal* 3(3): 295–302.

Roberts, Donald R., S. Manguin, and J. Mouchet. 2000. DDT House Spraying and Re-emerging Malaria. *Lancet* 356(9226): 330–32.

Robinson, Nigel J., Catherine M. Procter, Erin L. Connolly, and Mary Lou Guerinot. 1999. A Ferric-chelate Reductase for Iron Uptake from Soils. *Nature* 397(6721): 694–97.

Rockett, Ian R. H. 1994. Population and Health: An Introduction to Epidemiology. *Population Bulletin* 49(3): 1–48.

Rodale. 2000. Prevention Magazine's National Survey on Self-care Movement Reveals 158 Million Consumers Use Dietary Supplements for Their Health and Spend $8.5 Billion Each Year Survey Also Reports That Widespread Use of Dietary Supplements May Cause Public Health Problems. *Prevention Magazine* Rodale Inc. Press release, Emmaus, Pa., 23 February.

Rodricks, Joseph V. 1992. *Calculated Risk: Understanding the Toxicity and Human Health Risks of Chemicals in the Environment.* Cambridge: Cambridge University Press.

Rose, C. S., J. W. Martyny, L. S. Newman, D. K. Milton, T. E. King, Jr., J. L. Beebe, J. B. McCammon, R. Hoffman, and K. Kreiss. 1998. "Lifeguard Lung": Endemic Granulomatous Pneumonitis in an Indoor Swimming Pool. *American Journal of Public Health* 88(12): 1795–1800

Rosen, Joseph D. 1990. Much Ado about Alar. *Issues in Science and Technology* 7(1): 85–90.

Rosenberg, Nathan. 1982. *Inside the Black Box: Technology and Economics.* Cambridge: Cambridge University Press.

Rothman, Stanley, and S. Robert Lichter. 1996. Is Environmental Cancer a Political Disease? In *The Flight from Science and Reason*, edited by Paul R. Gross, Norman Levitt, and Martin W. Lewis. New York: The New York Academy of Sciences.

RS. 1999a. Reports of the Six Royal Society Referees. London: The Royal Society, 10 May.

———. 1999b. Review of Data on Possible Toxicity of GM Potatoes, London: The Royal Society, 18 May.

———. 2000. UK Needs to Be Vigilant on Gender-bending Chemicals. London: The Royal Society, 28 June.

RS et al. 2000. Transgenic Plants and World Agriculture: Report Prepared under the Auspices of the Royal Society of London, the U.S. National Academy of Sciences, the Brazilian Academy of Sciences, the Chinese Academy of Sciences, the Indian National Science Academy, the Mexican Academy of Sciences and the Third World Academy of Sciences. Washington, D.C.: National Academy Press, July.

Rubin, Charles T. 1994. *The Green Crusade: Rethinking the Roots of Environmentalism.* New York: The Free Press.

Ruschitzka, Frank, Peter J. Meier, Marko Turina, Thomas F. Luscher, and Georg Noll. 2000. Acute Heart Transplant Rejection Due to Saint John's Wort *Lancet* 355(9203): 548.

Ruskin, John. 1880. *The Seven Lamps of Architecture*. Boston: Aldine.

Russell, Christine. 1990. A Year after Alar: The Pesticide Scare Grows. *Health: A Weekly Journal of Medicine, Science and Society. Washington Post*, 27 February, pp. 12–15.

Rybczynski, Witold. 1983. *Taming the Tiger: The Struggle to Control Technology*. New York: Viking Press.

Sachs, Susan. 1998. Public Clamor Puts Focus on "Clusters" in Cancer Research: Public Anger Propels Cancer Studies Focusing on Pollutants, Despite Skepticism. *New York Times*, 21 September.

Safe, Stephen H. 1994. Dietary and Environmental Oestrogens and Anti-oestrogens and their Possible Role in Human Disease. *Environmental Science and Pollution Research* 1: 29–33.

———. 1995. Environmental and Dietary Oestrogens and Human Health: Is There a Problem. *Environmental Health Perspectives* 103: 346–51.

———. 1997a. Is There an Association between Exposure to Environmental Estrogens and Breast Cancer? *Environmental Health Perspectives* 105 Supplement 3, March.

———. 1997b. Xenoestrogens and Breast Cancer. *New England Journal of Medicine* 337(8): 1303–4.

———. 1998. Interactions between Hormones and Chemicals in Breast Cancer. *Annual Review of Pharmacology and Toxicology* 38: 121–58.

———. 2000. Endocrine Disruptors and Human Health—Is There a Problem? An Update. *Environmental Health Perspectives* 108(6).

Safe, Stephen H., and K. Gaido. 1998. Phytoestrogens and Anthropogenic Estrogenic Compounds. *Environmental Toxicology and Chemistry* 17(1): 119.

Safe, Stephen, Warren G. Foster, James C. Lamb, Retha R. Newbold, and Glen Van Der Kraak. 2000. Estrogenicity and Endocrine Disruption. Council for Agricultural Science and Technology (CAST), 16, July.

Sahai, Suman. 1997. The Bogus Debate on Bioethics. *Biotechnology and Develoment Monitor*, March.

———. 1999. Biotechnology Capacity of LDCs in the Asian Pacific Rim. *AgBioForum* 2(3&4): 189–97.

Salazar-Lindo, Eduardo, Marcos Alegre, Marcela Rodriguez, Pedro Carrion, and Nora Razzeto. 1993. The Peruvian Cholera Epidemic and the Role of Chlorination in Its Control and Prevention. In *Safety of Water Disinfection: Balancing Chemical and Microbial Risks*, edited by Gunther F. Cruan. Washington, D.C.: ILSI (International Life Sciences Institute) Press.

Salinger, Pierre. 1989. High and Mighty. *American Way* 15, April.

Sanders, T. A. B. 1995. Vegetarian Diets and Children. *Pediatric Clinics of North America* 42(4): 955–65.

Sanders, T. A. B., and S. Reddy. 1994. Vegetarian Diets and Children, *American Journal of Clinical Nutrition* 59(5): 1176–81s.

SANR 2000. Food Safety and Organic Products: Dispelling the Myths. Soil Association News Release (UK), 17 February.

Schecter, A., P. Toniolo, L. C. Dai, L. T. Thuy, and M. S. Wolff. (1997). Blood Levels of DDT and Breast Cancer Risk among Women Living in the North of Vietnam. *Archives of Environmental Contamination and Toxicology* 33(4): 453–56.

Schick, Kathy D., and Nicholas Toth. 1993. *Making Silent Stones Speak: Human Evolution and the Dawn of Technology.* New York: Simon & Schuster.

Schiffer, Michael B. 1992. *Technological Perspectives on Behavioral Change.* Tucson: University of Arizona Press.

———. 1999a. Commentary: Return to Holism. *Anthropology Newsletter* 40(3): 62, 64.

———. 1999b. *The Material Life of Human Beings: Artifacts, Behavior, and Communication.* New York: Routledge.

Schiller, Friedrich. 1954. *On the Aesthetic Education of Man.* Trans. R. Snell. New York: Ungar.

Schilling, Govert. 2000. The Virtual Observatory Moves Closer to Reality. *Science* 289(5477): 238–39.

Schull, W. J. 1990. *Song among the Ruins.* Cambridge: Harvard University.

Schwab, Ivan R., and R. Rivkah Isseroff. 2000a. Bioengineered Corneas—The Promise and the Challenge. *New England Journal of Medicine* 343(2).

———. 2000b. Successful Transplantation of Bioengineered Tissue Replacement in Patients with Ocular Surface Disease. *Cornea: The Journal of Cornea and External Diseases* 19(4): 421–26.

ScienceScope. 1999. Food Fight. *Science* 283(5406).

Scobie, Grant M. 1979. *Investment in International Agricultural Research: Economic Dimensions.* World Bank Working Paper 361. Washington, D.C.: World Bank.

Sebestik, Jan. 1983. The Rise of Technological Science. *Historical Technology* 1(1): 25–43.

Serageldin, Ismail. 1999a. Panelist's Remarks. In *Biotechnology and Biosafety: A Forum,* edited by Ismail Serageldin and Wanda W. Collins. Washington, D.C.: World Bank.

———. 1999b. Biotechnology and Food Security in the 21st Century. *Science* 285(5426): 387–89.

Servat, E., J.-M. Lapetite, J.-C. Bader, and J.-F. Boyer. 1990. Satellite Data Transmission and Hydrological Forecasting in the Fight against Onchocerciasis in West Africa. *Journal of Hydrology* 117(1-4): 187–98.

Sharma, Rohit. 2000. India Eradicates Guinea Worm Disease. *BMJ (British Medical Journal)* 320(7236): 668.

Shelton, Antony M., and Richard T. Roush. 1999. False Reports and the Ears of Men. *Nature Biotechnology* 17(9): 832.

Shorter, Edward. 1982. *A History of Women's Bodies: A Social History of Women's Encounter with Health, Ill-Health, and Medicine.* New York: Basic Books.

———. 1991. *Women's Bodies: A Social History of Women's Encounter with Health, Ill-Health, and Medicine.* New Brunswick, N.J.: Transaction Publishers.

Siddiqui, Bahar A., and Samiullah Khan, eds. 1999. *Breeding in Crop Plants: Mutations and In Vitro Mutation Breeding.* New Delhi: Kalyani Publishers.

Siddle, Julian. 2000. Move to Control Pesticides. BBC World Service online, 3 December.

Sidley, Pat. 2000. Malaria Epidemic Expected in Mozambique. *BMJ British Medical Journal)* 320(7236): 669.

Simberloff, Daniel, and Peter Stiling. 1996a. Risks of Species Introduced for Biological Control. *Biological Conservation* 78(1/2): 185–92.

———. 1996b. How Risky Is Biological Control? *Ecology* 77: 1965–74.

———. 1998. How Risky Is Biological Control? Reply. *Ecology* 79(5): 1834–36.

Singleton, Janet. 1999. Herbal Remedy Rapped—Misconceptions about Medications—Two Articles. *Denver Post*, 4 March.

Slakey, Francis 1993. When the Lights of Reason Go Out. *New Scientist.* 139(1890): 49–50.

Smil, Vaclav. 1991. The Critical Link between Population Growth and Nitrogen. *Population and Development Review* 17(4): 569–601.

———. 1993. *Global Ecology: Environmental Change and Social Flexibility.* New York: Routledge.

Smith, A. G. 2000. How Toxic Is DDT? *Lancet* 356(9226): 267–68.

Smith, Lisa C., and Lawrence Haddad. 2000. *Overcoming Child Malnutrition in Developing Countries: Past Achievements and Future Choices.* IFPRI (International Food Policy Research Institute) 2020 Vision for Food, Agriculture, and the Environment Discussion Papers.

Smith, T. L., M. L. Pearson, K. R. Wilcox, C. Cruz, M. V. Lancaster, B. Robinson-Dunn, F. C. Tenover, M. J. Zervos, J. D. Band, E. White, and W. R. Jarvis. 1999. Emergence of Vancomycin Resistance in *Staphylococcus aureus. New England Journal of Medicine* 340(7).

Solman, Paul. 1999. World Bank Doubts Prices Recovery. *Financial Times (London)*, 3 February.

Sopher, David E. 1979. The Landscape of Home: Myth, Experience, Social Meaning. In *The Interpretation of Ordinary Landscapes: Geographical Essays*, edited by D. W. Meinig. New York: Oxford University Press.

Spindler, M. 1983. DDT: Health Aspects in Relation to Man and Risk/benefit Assessment Based Thereupon. *Residue Reviews*, 90.

Squire, G. R., N. Augustin, J. Brown, J. W. Crawford, G. Dunlop, J. Graham, J. R. Hillman, B. Marshall, D. Marshall, G. Ramsay, D. J. Robinson, J. Rissell, C. Thompson, and G. Wright. 2000. Gene Flow in the Environment-Genetic pollution? *SCRI (Scottish Crop Research Institute) 1998/99 Annual Report* 45-54, Invergowie, Dundee, Scotland, January.

Stanford, Craig B. 1995a. To Catch a Colobus. *Natural History* 104(1): 48.

———. 1995b. Chimpanzee Hunting Behavior-Chimpanzees Use Meat as a Political and Reproductive Tool. *American Scientist* 83(3): 256–61.

———. 1996a. The Hunting Ecology of Wild Chimpanzees: Implications for the Evolutionary Ecology of Pliocene Hominids. *American Anthropologist* 98(1): 96–113 .

———. 1996b. The Colobines: Beyond Infanticide—A Book Review. *American Journal of Primatology* 38(2): 187–89.

———. 1998a. Predation and Male Bonds in Primate Societies. *Behaviour* 135(4): 513–33.

———. 1998b. *Chimpanzee and Red Colobus: The Ecology of Predator and Prey.* Cambridge: Harvard University Press.

———. 1999. *The Hunting Apes: Meat Eating and the Origins of Human Behavior.* Princeton: Princeton University Press.

Steele, James H. 1999. Food Irradiation: A Lost Public Health Opportunity. Minnesota Health Department, Irradiated Food Conference, Minneapolis, Minnesota, 21–22 June.

Steiner, Rudolf. 1958. *Agriculture: A Course of Eight Lectures.* Trans. George Adams. London: Biodynamics Association.

Stellman, S. D., M. J. Djordjevic, J. A. Britton, J. E. Muscat, M. L. Citron, M. Kemeny, E. Busch, and L. Gong. 2000. Breast Cancer Risk in Relation to Adipose Concentrations of Organochlorine Pesticides and Polychlorinated Biphenyls in Long Island, New York. *Cancer Epidemiology, Biomarkers & Prevention* 9(11): 1241–49.

Stephenson, Joan. 1997a. Public Health Experts Take Aim at a Moving Target: Foodborne Infections. *JAMA (Journal of the American Medical Association)* 277(17): 97–98.

———. 1997b. New Approaches for Detecting and Curtailing Foodborne Microbial Infections. *JAMA (Journal of the American Medical Association)* 277(17): 1337–40.

———. 2000. Easy on the Antioxidants. *JAMA (Journal of the American Medical Association)* 283(19).

STN. 1999. Survey (by Eurobarometer and Thomas J. Hoban), *Seed Trade News* (Ball Publishing, Batavia, Ill.), December.

Stokstad, Erik. 1999a. Animal Experimentation: Humane Science Finds Sharper and Kinder Tools. *Science* 286(5442): 1068–71.

————. 1999b. Animal Experimentation: Toxicity Testing: The Many Arts of Persuasion. *Science* 286(5442): 1070.

Stone, Richard. 1994. Environmental Estrogens Stir Debate. *Science* 265(5170): 308–10.

————. 1995. A Molecular Approach to Cancer Risk. 268(5209): 356–57.

Strachan D. P. 1989. Hay Fever, Hygiene, and Household Size. *BMJ (British Medical Journal)*299(6710): 1259–60.

Strawley, George. 2000. Study Finds No Link between Three Mile Island Accident, Cancer Deaths. Associated Press, Nando Media online, 27 April.

Swaminathan, M. S. 2000a. Science in Response to Basic Human Needs. *Science* 287(5452): 425.

————. 2000b. Genetic Engineering and Food Security: Ecological and Livelihood Issues. In *Agricultural Biotechnology and the Poor*, edited by Gabrielle J. Persley and M. M. Lantin. Washington, D.C.: Consultitative Group on International Agricultural Research and the U. S. National Academy of Sciences.

Symons, James M. 1992. *Plain Talk about Drinking Water*. Boulder: American Waterworks Association.

————. 1995. *Drinking Water: Refreshing Answers to All Your Questions*. College Station: Texas A&M University Press.

Tacket, Carol O., Hugh S. Mason, Genevieve Losonsky, Mary K. Estes, Myron Levine, and Charles Arntzen. 2000. Human Immune Responses to a Novel Norwalk Virus Vaccine Delivered in Transgenic Potatoes. *The Journal of Infectious Diseases* 182(1).

Talbott, Evelyn O., Ada O. Youk, Kathleen P. McHugh, Jeffrey D. Shire, Aimin Zhang, Brian P. Murphy, and Richard A. Engberg. 2000. Mortality among the Residents of the Three Mile Island Accident Area: 1979–1992. *Environmental Health Perspectives* 108(6): 545–52 (posted online 27 April).

Taormina, Peter J., Larry R. Beuchat, and Laurence Slutsker. 1999. Infections Associated with Eating Seed Sprouts: An International Concern. *Emerging Infectious Diseases Journal* 5(5).

Tauxe, Robert V. 1991. Salmonella: A Postmodern Pathogen. *Journal of Food Protection* 54: 563–68.

————. 1997a. Emerging Foodborne Diseases: An Evolving Public Health Challenge. *Emerging Infectious Diseases Journal* 3(4): 425–34.

————. 1997b. Does Organic Gardening Foster Foodborne Pathogens: In Reply. *JAMA (Journal of the American Medical Association)* 277(21): 1680.

————. 1998. Foodborne Illnesses. Strategies for Surveillance and Prevention. *Lancet* 352(9145).

Taverne, Dick. 1999. Against Anti-science. *Prospect* 47, December.

Taylor, Brent, Elizabeth Miller, C. Paddy Farrington, Maria-Christina Petropoulos, Isabelle Favot-Mayaud, Jun Li, and Pauline A. Waight.

1999. Negative Association between MMR and Autism. *Lancet* 353(9169), 12 June.

Tenner, Edward. 1996. *Why Things Bite Back: Technology and the Revenge of Unintended Consequences.* New York: Knopf.

Thayer, Ann M. 1999. Transforming Agriculture: Transgenic Crops and the Application of Discovery Technologies Are Altering the Agrochemical and Agriculture Businesses. *Chemical & Engineering News* 77(16): 21–35.

Thompson, Kimberly M. 2000. Kids at Risk. *Risk in Perspective: An Official Publication of the Harvard Center for Risk Analysis* 8(4): 1–6.

Thompson, Larry. 2000. Are Bioengineered Foods Safe? *FDA Consumer Magazine (Food and Drug Administration)*, January–February.

Thompson, Lea. 1998. A New Name Game: What's in Your Water's Origins. Dateline NBC online, 2 September.

Thompson, Sylvanus, and D. A. Powell. 2000. Risks Associated with the Consumption of Fresh Sprouts, Food Safety Network Technical Report No. 16, July.

Tierney, John. 1999. The Big City: A Tale of Public Perceptions, Global Politics and Mosquito Killers. *New York Times*, 16 September.

Tomlinson, John. 1991. *Cultural Imperialism.* Baltimore: Johns Hopkins University Press.

Tren, Richard. 2000. After the Floods: Malaria Will Now Take Its Toll in Mozambique. *Wall Street Journal* (Europe), 6 March.

Tren, Richard, and Donald R. Roberts. 1999. The Economic Costs of Malaria in South Africa: Malaria Control and the DDT Issue. IEA (The Institute of Economic Affairs) online, August.

Trewavas, Anthony. 1999. Much Food, Many Problems. *Nature* 402(6759): 231.

———. 2000. Toxins and Genetically Modified Food. *Lancet* 355(9207): 931–34.

Trewavas, Anthony, and C. J. Leaver. 1999. Conventional Crops Are the Test of GM Prejudice. *Nature* 401(6754): 640.

TRF. 1999a. New Rices May Help Address Vitamin A and Iron Deficiency, Major Causes of Death in the Developing World. Press release, The Rockefeller Foundation, 3 August.

———. 1999b. Breakthrough in Rockefeller-Funded Research to Combat Nutritional Deficiencies Announced. *Philanthropy News Digest*, 10 August.

Tsai, Ray Jui-Fang, Lien-Min Lia, and Jan-Kan Chen. 2000. Reconstruction of Damaged Corneas by Transplantation of Autologous Limbal Epithelial Cells. *New England Journal of Medicine* 343(2).

Tuljapurkar, Shripad, Nan Li, and Carl Boe 2000. A Universal Pattern of Mortality Decline in the G7 Countries. *Nature* 405(6788): 789–92.

Turner, Terence. 1993. Anthropology and Muticulturalism: What Is Anthropology That Multiculturalists Should Be Mindful of It? *Cultural Anthropology: Journal of the Society for Cultural Anthropology* 8(4): 411–29.

Twitchell, James B. 1992. *Carnival Culture: The Trashing of Taste in America*. New York: Columbia University Press.

UNDP 1998. *Human Development Report: 1998*. New York: Oxford University Press for the United Nations Development Programme.

UNFPA. 1999. FAO Says 800 Million Sleep Hungry Each Night. *POPULI: The UNFPA Magazine*, United Nations Population Fund, December.

UNICEF. 1998. *The State of the World's Children 1998*. New York: Oxford University Press for United Nations Children's Fund.

UNSCEAR. 2000. UNSCEAR 2000 Report to the General Assembly. Vienna, Austria: United Nations Scientific Committee on the Effects of Atomic Radiation, 6 June.

USAID. 1998. Waging War against Infectious Diseases. *USAID Developments* 4(3): 1.

USC. 2000. Sperm Counts Unchanged over 50 Years. University of Southern California Press release, 28 March.

USFDA (United States Food and Drug Administration, Center for Food Safety and Applied Nutrition). 1991. Testimony before the Subcommittee on Human Resources and Intergovernmental Relations Committee on Governmental Operations, U.S. House of Representatives, Washington, D.C.,18 July 1991, (hypertext updated, 7 August 1997)

———. 1993. *Genetically Engineered Foods: Fears and Facts: An Interview with FDA's Jim Maryanski*. Washington, D.C.: U.S. Food and Drug Administration, January/February.

———. 2000. *Bt Corn Less Insect Damage, Lower Mycotoxin Levels, Healthier Corn*. Washington, D.C.: U.S. Food and Drug Administration, ARS News Service, 26 April.

USHR. 2000a. *Seeds of Opportunity: An Assessment of the Benefits, Safety, and Oversight of Plant Genomics and Agricultural Biotechnology*. Washington, D.C.: U.S. House of Representatives, Committee on Science, 13 April.

———. 2000b. *Smith Releases Report on Genetically-modified Plants*. Washington, D.C.: U.S. House of Representatives, Committee on Science, 13 April.

Van Beneden, C. A., W. E. Keene, R. A. Strang, D. H. Werker, A. S. King, B. Mahon, K. Hedberg, A. Bell, M. T. Kelly, V. K. Balan, W. R. Mac Kenzie, and D. Fleming. 1999. Multinational Outbreak of Salmonella enterica Serotype Newport Infections Due to Contaminated Alfalfa Sprouts. *JAMA (Journal of the American medical Association)* 281(2): 158–62.

van't Veer, P., I. E. Lobbezoo, J. M. Martin-Moreno, E. Gualla, J. Gomez-Ara-
     cena, A. F. Kardinaal, L. Kohlmeier, B. C. Martin, J. J. Strain, M.
     Thamm, P. van Zoonen, B. A. Baumann, J. K. Huttunen, and F. J. Kok.
     1997. DDT (dicophane) and Postmenopausal Breast Cancer in Europe:
     Case-control Study. *BMJ* 315(7100): 81–85.
Veblen, Thorstein. 1922. *The Instinct of Workmanship and the State of Indus-
     trial Arts.* New York: B.W. Huebsch.
———. 1934. *The Theory of the Leisure Class: An Economic Study of Insti-
     tutions.* New York: The Modern Library.
Vergano, Dan. 1999. Scientific Misconduct: EMF Researcher Made Up Data,
     ORI Says. *Science* 285(5424): 23–25.
Vidal, Aline. 1988. *Eiffel's Towering Achievement, France.* Washington, D.C.:
     La Maison Francaise.
Voelker, Rebecca. 2000a. The Hygiene Hypothesis. *JAMA (Journal of the
     American medical Association)* 283(10).
———. 2000b. Africa Needs More Aid, New Partnerships to Fight Malaria.
     JAMA (Journal of the American Medical Association) 283(19).
Vogel, David. 1997. *Trading Up: Consumer and Environmental Regulation in
     a Global Economy.* Cambridge: Harvard University Press.
von Schenck, Ursula, Christine Bender-Gotze, and Berthold Koletzko. (1997).
     Persistence of Neurological Damage Induced by Dietary Vitamin B-12
     Deficiency in Infancy. *Archives of Disease in Childhood* 77(2): 137–39.
Waage, Jeff K. 1995. Divergent Perspectives on the Future of IPM. Paper Pre-
     pared for the IFPRI Workshop Pest Management, Food Security and the
     Environment: The Future to 2020. Washington, D.C.: International
     Food Policy Research Institute (IFPRI), May.
Wakeford, Tom. 1991. A Green in the Machine: As a Government Scientist,
     Norman Moore Both Initiated the Research That Showed How Pesti-
     cides Damage Wildlife and Framed the Laws that Brought these Chem-
     icals Under Control. *New Scientist* 132(1743).
Waldvogel, Francis A. 1999. New Resistance in Staphylococcus Aureus. *New
     England Journal of Medicine* 340(7).
Wambugu, Florence. 1999. Why Africa Needs Biotech. *Nature* 400(6739):
     15–16.
———. 2000. Feeding Africa. *New Scientist* 165(2240): 40–43.
Warchol, Alice. 1998. "Natural" Supplements—Safe, or Just Untested?
     Scripps-Howard, Nando.net online, 25 April.
Washburn, Sherwood. 1960. Tools and Human Evolution. *Scientific American*
     203(3): 62–75.
Weber, Jocelyne R. 2000. Review of Agricultural Biotechnology in Africa.
     AfricaBio www.up.ac.za/academic/fabi/africabio/.
Weiderpass, E., T. Partanen, R. Kaaks, H. Vainio, M. Porta, T. Kauppinen, A.
     Ojajarvi, P. Boffetta, and N. Malats. 1998. Occurrence, Trends and En-

vironmental Etiology of Pancreatic Cancer. *Scandanavian Journal of Work and Environmental Health* 24(3): 165–74.

Wellness Letter. 1998. Water in the Wild. *University of California at Berkeley Wellness Letter: The Newsletter of Nutrition, Fitness, and Stress Management* 14(3): 4–5.

Wessely, Simon. 2000. Responding to Mass Psychogenic Illness. *New England Journal of Medicine* 342(2).

Westendorp, R. G. J., and T. B. L. Kirkwood. 1998. Human Longevity at the Cost of Reproductive Success. *Nature* 396(6713): 743–46.

Whelan, Elizabeth M. 1993. *Toxic Terror: The Truth behind the Cancer Scares*. New York: Prometheus Books.

———. 2000. An "Alice-in-Wonderland" Approach to Disease Prevention. American Council on Science and Health editorial online, 26 June.

White, Kevin M., and Samuel H. Preston. 1996. How Many Americans Are Alive Because of Twentieth-century Improvements in Mortality? *Population and Development Review* 22(3): 415–29.

White, Lon R., Helen Petrovitch, Webster Ross, Kamal Masaki, John Hardman, James Nelson, Daron Davis, and William Markesbery. 2000. Brain Aging and Midlife Tofu Consumption. *Journal of the American College of Nutrition* 19(2): 242–55.

White, Randall. 1986. *Dark Caves, Bright Visions: Life in Ice Age Europe*. New York: W.W. Norton.

Whitman, David 1998. *The Optimism Gap: The I'm Ok, They're Not Syndrome and the Myth of American Decline*. New York: Walker and Company.

Whitman, Lois. 2000. Abusive Child Labor Found in U.S. Agriculture: U.S. Law Discriminates against Child Farmworkers, Children's Rights Division of Human Rights Watch, Press release, 20 June.

WHO. 1971. The Place of DDT in Operations against Malaria and Other Vector-borne Diseases. *WHO Official Records,* Executive Board, Forty-Seventh Session, Part II, Appendix, 190, 176–82.

———. 1979. *DDT and Its Derivatives Published Under the Joint Sponsorship of the United Nations Environment Programme and the World Health Organization*. Environmental Health Criteria 9, WHO Task Group on Environmental Health Criteria for DDT and its Derivatives. Geneva: World Health Organization.

———. 1995. *Onchocerciasis and Its Control: Report of a WHO Expert Committee on Onchocerciasis Control*. Geneva: World Health Organization, WHO Technical Report Series No. 852.

———. 1996. Onchocerciasis: Progress Towards Elimination in the Americas. *Weekly Epidemiological Record (World Health Organization)* 71(37): 227–29.

———. 1997. Health and Sustainable Development: 5 Years after the Earth Summit. Geneva: Press Release WHO/47, 18 June.

———. 1998. 1998 *World Health Report: Life in the 21st Century: A Vision for All.* Geneva: World Health Organization.

———. 1999a. *World Health Assembly Concerned about Persistent Organic Pollutants.* Geneva: World Health Organization.

———. 1999b. *Arsenic in Drinking Water,* Fact Sheet No 210. Geneva: World Health Organization.

———. 1999c. *Onchocerciasis Contol Programme in West Africa: In a Nutshell.* Geneva: World Health Organization.

———. 1999d. *Building Partnerships for Lymphatic Filariasis: Strategic Plan.* Geneva: World Health Organization.

———. 1999e. *Bacillus thuringiensis.* Environmental Health Criteria No. 217. Geneva: World Health Organization.

———. 2000a. *Mozambican Floods Leave 250,000 People at Risk from Disease.* Geneva: World Health Organization, 13 March.

———. 2000b. *Time Limited Exemptions and Financial Support Are Critical to Sustainable Reductions in the Use of DDT.* Geneva: World Health Organization, 17 March.

———. 2000c. The Bill and Melinda Gates Foundation Announces New Global Health Grant to Accelerate the Campaign to Eradicate Guinea Worm Disease. Geneva: World Health Organization Press release.

———. 2000d. WHO Issues New Healthy Life Expectancy Rankings. Geneva: World Health Organization Press release, 4 June.

———. 2000e. *World Health Report 2000.* Geneva: World Health Organization.

Whorton, James C. 1974. *Before Silent Spring: Pesticides and Public Health in Pre-DDT America.* Princeton: Princeton University Press.

Wilairatana, Polrat, and Sornchai Looareesuwan. 2000. Breathing in the Tropics Made Easy. *Lancet* 355(9209): 1108.

Willett, John 1984. *The Weimar Years: A Culture Cut Short.* New York: Thames and Hudson.

Williams, Francis. 1999. Hormones Add to Pressure on WTO. *Financial Times (London),* 2 February.

Williamson, David. 2000. Study: Commercial Disinfectants Effective, Natural Products Less So. Press release, University of North Carolina at Chapel Hill, 24 January.

Willis, Paul E. 1990. *Common Culture: Symbolic Work at Play in the Everyday Cultures of the Young.* Boulder: Westview.

Wilson, Frank R. 1998. *The Hand: How Its Use Shapes the Brain, Language, and Human Culture.* New York: Pantheon Books.

Wilson, Michael, John R. Hillman, and David J. Robinson. 1999. Genetic Modification in Context and Perspective. In *Fearing Food: Risk, Health and Environment,* edited by Julian Morris and Roger Bate. Boston: Butterworth-Heinemann.

Windschuttle, Keith 1997. *The Killing of History: How Literary Critics and Social Theorists Are Murdering Our Past.* New York: The Free Press.

Winestock, Geoff. 2000. EU Beef Battle with U.S. Began in a Rare Event. Dow Jones online, 2 March.

Winslow, Lisa Corbin. 1998. Herbal Medicine. *Cleveland Clinic Journal of Medicine* 65(7): 388–89.

Winslow, Lisa Corbin, and D. J. Kroll. 1998. Herbs As Medicine. *Archives of Internal Medicine* 158(21): 2192–99.

Wong, Craig S., Srdjan Jelacic, Rebecca L. Habeeb, Sandra L. Watkins, and Phillip I. Tarr. 2000. The Risk of the Hemolytic-Uremic Syndrome after Antibiotic Treatment of *Escherichia coli* O157:H7 Infections. *New England Journal of Medicine* 325(25), June 29 (Notice posted 23 May 2000).

Wood, David. 1998. Ecological Principles in Agricultural Policy: But Which Principles? *Food Policy* 23(5): 371–81.

Woodward, Calvin. 1997. Bottled Water: Tastes Great, But Does It Mean More Fillings? Associated Press Nando.net online, 14 July.

World Bank. 1993. *World Development Report 1993: Investing in Health.* New York: Oxford University Press.

———. 1998. *World Development Report 1998/99: Knowledge for Development.* New York: Oxford University Press.

Wouters, Inge M., Jeroen Douwes, Gert Doekes, Peter S. Thorne, Bert Brunekreef, and Dick J. J. Heederik. 2000. Increased Levels of Markers of Microbial Exposure in Homes with Indoor Storage of Organic Household Waste. *AEM (Applied and Environmental Microbiology)* 66(2): 627–31.

Wraight, C. L., A. R. Zangerl, M. J. Carroll, and M. R. Berenbaum. 2000. Absence of Toxicity of *Bacillus thuringiensis* Pollen to Black Swallowtails under Field Conditions. *PNAS (Proceedings of the National Academy of Sciences),* 6 June.

WRI. 1998. *World Resources 1998–99 A Guide to the Global Environment: People and the Environment: Environmental Change and Human Health.* New York: Oxford University Press for the World Resources Institute.

Wrong, Michela. 2000. Gordon Conway: An Advocate of Cautious Change. *Financial Times (London),* 5 May.

WWF. 1998. Time to Ban DDT in Africa, Says WWF. *Electronic Mail & Guardian* online, 30 June.

———. 1999a. Governments Losing Focus on Elimination of World's Deadliest Chemicals. *WWF International,* 29 January.

———. 1999b. A WWF Issue Brief: Persistent Organic Pollutants: Hand Me-down Poisons That Threaten Wildlife and People. *WWF International.*

———. 1999c. WWF Report Finds Up to 82 Percent of DDT Sprayed Indoors Escapes into Environment. *WWF International.*

————. 1999d. Resolving the DDT Dilemma: Protecting Human Health and Biodiversity. *WWF International.*

Yamey, Gavin. 2000a. Global Alliance Launches Plan to Eliminate Lymphatic Filariasis. *BMJ (British Medical Journal)* 320(7230): 269.

————. 2000b. African Heads of State Promise Action against Malaria. *BMJ (British Medical Journal)* 320(7244): 1228.

Yao, Shujie. 2000. Economic Development and Poverty Reduction in China over 20 Years of Reform. *Economic Development and Cultural Change* 48(3): 447–74.

Ye, Xudong, Salim Al-Babili, Andreas Kloti, Jing Zhang, Paola Lucca, Peter Beyer, and Ingo Potrykus. 2000. Engineering the Provitamin A (Beta-Carotene) Biosynthetic Pathway into (Carotenoid-Free) Rice Endosperm. *Science* 287(5451): 303–5.

YCC. 2000. No Link Found between PBCs and DDE and Breast Cancer Risk in Women. New Haven: Yale Cancer Center Press releases, 18 February.

Yoon, Carol Kaesuk. 1997. Tuning in to the Hidden Message in Some Fruit Siren Songs: Toxins May Ward Off Fungi So Seeds Can Be Dispersed. *New York Times*, 8 April.

Yudelman, Montague, Annu Ratta, and David Nygaard. 1998. *Pest Management and Food Production.* Food Agriculture and the Environment Discussion Paper 25. Washington, D.C.: International Food Policy Research Institute.

Yue, Qun-Ying, Carin Bergquist, and Barbro Gerden. 2000. Seven Cases of Decreased Effect of Warfarin during Concomitant Treatment with St. John's Wort. *Lancet* 355(9203): 576.

Zheng, T., T. R. Holford, S. T. Mayne, B. Ward, D. Carte, P. H. Owens, R. Dubrow, S. H. Zahm, P. Boyle, S. Archibeque, and J. Tessari. 1999. DDE and DDT in Breast Adipose Tissue and Risk of Female Breast Cancer. *American Journal of Epidemiology* 150(5): 453–58.

Zheng, T., T. R. Holford, S. T. Mayne, J. Tessari, B. Ward, D. Carter, P. H. Owens, P. Boyle, R. Dubrow, S. Archibeque-Engle, O. Dawood, and S. H. Zahm. 2000. Risk of Female Breast Cancer Associated with Serum Polychlorinated Biphenyls and 1,1-Dichloro-2,2'-bis (p-chlorophenyl)ethylene1. *Cancer Epidemiology, Biomarkers & Prevention* 9(2).

Zink, Don L. 1997. The Impact of Consumer Demands and Trends on Food Processing. *Emerging Infectious Diseases Journal* 3(4).

Index

n after a number refers to a note.